Guide to TV and Video Technol

DATE DUE

AUG 5 08			

DEMCO 38-296

Guide to TV and Video Technology

Second edition

Eugene Trundle, TMIEEIE, MRTS, MISTC

Newnes

OXFORD   AUCKLAND   BOSTON   JOHANNESBURG   MELBOURNE   NEW DELHI

3DP
301-2041
ssional Publishing Ltd

A member of the Reed Elsevier plc group

First published 1988 as *Newnes Guide to TV and Video Technology*
Reprinted 1990, 1992
Second edition 1996
Reprinted 1997, 1999

© Eugene Trundle 1988, 1996

British Library Cataloguing in Publication Data
A catalogue record for this book is available from the British Library

Library of Congress Cataloguing in Publication Data
A catalogue record for this book is available from the Library of Congress

ISBN 0 7506 23748

Typeset by Keyword Typesetting Services Ltd, Wallington, Surrey
Printed and bound in Great Britain by MPG Books Ltd, Bodmin, Cornwall

Contents

Preface

The terms of reference for this book are very wide, and increase with each new edition. TV and video take an ever larger part in our leisure, educational and recreational activities; here the technology behind them is explored and explained.

Much use is made throughout the book of block diagrams. Since integrated circuits – silicon chips – are so widespread, much service data are now presented in block diagram form. To explain principles and techniques I have often used early 'discrete' circuits, in which the separate functions and circuit elements can be clearly discerned.

This book is aimed at interested laypeople, students, technicians and those in allied fields seeking an insight into the technicalities of TV and VTR practice. I have assumed that the reader has a basic knowledge of electronics and mechanics. Although the book addresses itself to domestic products, the line between these and professional gear is now very blurred and ill-defined – the same techniques and even components are used in both these days! Similarly, the various VTR formats are amazingly similar in essence.

For further reading I can recommend my *Television and Video Engineer's Pocket Book* and Steve Beeching's *Videocassette Recorders – A Servicing Guide*, both published by Butterworth-Heinemann. Manufacturers' service manuals and technical training courses are a mine of product-specific information.

My thanks are due to several setmakers for their help and for supplying pictures, individually acknowledged in the captions; to the Engineering Information Departments of the BBC and ITC; and to my patient and loving wife Anne without whose moral support, coffee-brewing and typing services I would have been open-circuited . . . . .

E. Trundle

1 Basic television

For a reasonable understanding of colour television, it is essential that the basic principles of monochrome TV are known. As we shall see, all colour systems are firmly based on the original 'electronic-image dissection' idea which goes back to EMI in the 1930s, and is merely an extension (albeit an elaborate one) of that system.

Although there are few black and white TVs or systems now left in use, the compatible colour TV system used today by all terrestrial transmitters grew out of the earlier monochrome formats. In the early days it was essential that existing receivers showed a good black and white picture from the new colour transmissions, and the scanning standards, *luminance* signal, and modulation system are the same. What follows is a brief recap of *basic* television as a building block of the colour TV system to be described in later chapters.

Image analysis

Because a picture has two dimensions it is only possible to transmit all the information contained within it in *serial* form, if we are to use but one wire or RF channel to carry the signal. This implies a *dissection* process, and requires a timing element to define the rate of analysis; this timing element must be present at both sending and receiving ends so that the analysis of the image at the sending end, and the simultaneous build-up of the picture at the receiver, occur in synchronism. Thus a television picture may be dissected in any manner, provided that the receiver assembles its picture in precisely the same way; but the path link between sender and viewer must contain *two* distinct information streams: video signal, which is an electrical analogy of the light pattern being sent, and timing signals, or synchronisation pulses, to define the steps in the dissection process. The presence of a timing element suggests that each picture will take a certain period to be built up; how long will depend on how quickly we can serialise the picture *elements*, and this in turn depends on the bandwidth available in the transmission system – more of this later.

Figure 1.1 *The scanning process. Horizontal lines are drawn from left to right of the screen by horizontal direction, and 'stacked' vertically by the slower-moving vertical deflection field*

Scanning

If we focus the image to be televised on a light-sensitive surface we are ready for the next stage in the dissection process – the division of the pattern into picture elements or *pixels*. Each pixel is rather like the individual dots that go to make up a newspaper photograph in that each can only convey one level of shading. Thus the detail, or definition, in the reproduced picture is proportional to the number of pixels. In 625-line television we have approximately 450 000 pixels, adequate for a 67 cm-diagonal picture, but barely sufficient for much larger screens. These individual pixels are arranged in horizontal lines; there are 625 lines in the British TV system. Figure 1.1 shows how the image is scanned, line by line, to read out in serial form the pattern of light and shade which forms the picture. When half the lines have been traced out the scanning spot has reached the bottom of the picture and traced one *field*. It now flies back to the top of the screen to trace out the rest of the 625 lines in the spaces between those of its first descent. This is known as interlacing, and confers the advantages of a 50 Hz (Hz, Hertz, one cycle per second) flicker rate with the lower scanning speed and lesser bandwidth requirement of a 25 Hz *frame* rate. All TV systems use this 2:1 interlaced field technique; its success depends only on accurate triggering of the field scan.

The camera tube

In practice the image is focused on the internal *target* behind the faceplate of a camera tube, which is usually a variant of the *vidicon* type, illustrated in Figure 1.2. The vidicon has a heated cathode at its back end, in similar style to a picture tube, and this emits a beam of electrons forward towards the target area. A system of electrodes within the tube performs the functions of beam accelerator and electron-lens, so that the beam is brought to pinpoint focus at the target surface. The scanning action is achieved by suitably-shaped waveforms in the line- and field-scanning coils surrounding the tube. These magnetically deflect the electron beam on its way along the tube so that the interlaced scanning pattern of Figure 1.1 is traced out on the target. To achieve the very small spot size and precise geometry required of the tiny scanning field, a magnetic focusing system is used in

Figure 1.2 *Primary elements of photoconductive vidicon camera tube*

conjunction with the electrostatic lens within the tube, and beam focus optimised by adjusting the DC current in the large cylindrical focus coil. Having set up our image-scanning system, we must now see how the image is read out as an electrical waveform.

Target operation

Figure 1.3 represents the equivalent electrical circuit of the target of a photoconductive camera tube. The target consists of a rectangular plate coated with a photoconductive material such as lead oxide whose electrical resistance in the dark is very high, falling to a low level when light is present on its surface. When a complex pattern of light and shade is present, in sharp focus, each picture element

Figure 1.3 *Equivalent circuit of photoconductive camera tube, showing the effective RC elements (see text)*

3

Figure 1.4 *Basic principle of scanning at the camera tube*

will effectively form a parallel RC pair, as in Figure 1.3, and the charge on the capacitor (formed between the front and back layers of the target coating) will be proportional to the light intensity at that point. The scanning beam is in effect a conductor, and as it periodically and momentarily picks out each pixel, the capacitor will charge to tube-cathode potential via the electron beam which acts as a conductor. Current for this charging process must come from the target voltage source, so that if we interpose a resistor in series with the positive target potential, we shall develop across it a signal voltage proportional to target current. This is our output signal, an electrical facsimile of the picture element brightness along each line as scanned by the beam. The principle is illustrated in Figure 1.4.

Scanning waveforms

The beam in the camera tube is deflected from right to left of the target (as viewed from within) and its passage takes just 52μs (μs, microsecond, 0.000001 second) during which it passes across the target at a constant velocity. Having reached the left-hand side, the beam flies back to its starting point on the right, taking about 12μs to do so. While this horizontal scanning process is taking place, the beam is simultaneously being slowly drawn downwards (by the action of the field deflection coil) until it reaches the bottom of the target after 20 ms (ms, millisecond, 0.001 second) whereupon it is suddenly deflected back to the top (field flyback or retrace) to begin the next sweep. Thus the current waveforms in the deflection coils need to be ramp- or sawtooth-shaped with very linear characteristics, and

4

Figure 1.5 *Comparison between the time scales of line and field scanning*

there needs to be $312\frac{1}{2}$ horizontal sweeps (half the total) to each vertical sweep, as in Figure 1.5.

The video signal

The waveform coming from the vidicon target will be similar to that shown in Figure 1.6, consisting of an *analogue* signal representing the picture pattern, with 'blanks' at 64 µs and 20 ms intervals, during which flyback or *retrace* is taking

Figure 1.6 *Lines of video signal: (a) high brightness, (b) low brightness, (c) picture content*

5

place. In practice there will be spurious signals during these *blanking* intervals, and these are suppressed in the camera circuit.

Charge-coupled devices (CCDs)

An alternative method of televising an image is now commonly used, using a semiconductor pickup device (sometimes called a solid-state image sensor) in place of the traditional thermionic tube. A very small (typically 8 × 10 mm) sensor 'chip' contains a matrix of many thousands of tiny silicon photodiodes, each of which effectively forms a capacitor whose charge is proportional to the brightness of light falling on it. These capacitors are represented by C1, C2, C3 and C4 in Figure 1.7, which portrays the first four pixels in one television line. C1 will have a charge proportional to the light level on the first pixel, and this will appear as a voltage at the output of the first amplifier A1. If all the switches S1 to S3 are now momentarily closed, the charge on C1 will be passed into C2, whose charge will be transferred into C3 and so on, all the way to the last in the line, at the right-hand side of the diagram. Thus, by momentarily closing all the switches in synchronism, the brightness level charge for each pixel can be made to 'march' like a column of soldiers to the right where an orderly sequential readout of the light pattern along that line can be picked up, at a rate dependent on the switching frequency. This will form the video signal, with each TV line being read-out in sequence, in similar fashion to a computer's *shift-register*. This is the basis of the operation of a CCD.

Figure 1.7 *The concept of CCD image sensor. Light patterns are held as charges in the capacitors, one for each pixel*

We now have one of our information streams (the video signal), and it's time to insert the second stream (timing pulses) to synchronise the TV receiver. For simplicity, we shall assume that our signal source is a vidicon tube.

Synchronisation pulses

The receiver or monitor which we shall use to display the picture has scanning waveform generators too, and these must run in perfect synchronism with those at work in the camera. This will ensure that each picture element picked up from the vidicon target is reproduced in the right place on the display. Plainly, if the camera sees a spot of light in the top right-hand corner of the picture, and the monitor's scanning spot is in the middle of the screen when it reproduces the light, the picture is going to be jumbled up!

This is prevented by inserting synchronising pulses (sync pulses for short) into the video waveform at regular intervals, and with some distinguishing feature to enable the TV monitor to pick them out. To signal the beginning of a new line scan, we insert a 4.7 µs negative-going pulse into each line blanking period, and to initiate field flyback a *series* of similar, closely-spaced pulses are inserted into each field blanking period. These are shown in Figure 1.8, which represents what is called a VBS (video, black-level and sync) or *composite* video signal. Black level is established at 0.3 V (300 mV) from which the signal rises to 1 V for peak white with lesser brightness levels giving correspondingly lower voltage. Each time a sync pulse occurs, the signal voltage drops to zero for its duration. The timing of the field sync-pulse trains is very critical for good interlace in the displayed picture, and the sync pulse generator is carefully designed to achieve this. The short period preceding the line sync pulse is called the *front porch*, and the rather longer (5.8 µs) post sync-pulse period is termed the *back porch*. The time spent on porches and sync pulse is known as the *blanking period*, and it's 12 µs between lines, and 1.6 ms between fields, as shown in Figure 1.8. The lower section of the diagram indicates the relationship between sync pulses and scanning current for both camera tube and picture tube.

Picture reproduction

We have now obtained a composite video signal, in a form which conveys both video and timing information. Let us now see how it is used to recreate an image on the screen of a picture-tube. For simplicity, we will assume we have a closed-circuit set-up, and that the camera and monitor are linked by a single coaxial cable. Figure 1.9 shows the arrangement. Here we have, at the sending end, a vidicon tube with the necessary lenses, scan coils and power supplies.

Figure 1.8 *The basic analogue TV signal – arrangement of video and sync information and the relationship between signal timing and scanning waveforms*

Figure 1.9 *A closed-circuit set-up showing derivation and use of the waveforms in Figure 1.8*

Attendant on it is a master sync pulse generator which triggers the sawtooth generators in the camera and also provides a blanking signal for use in the video processing amplifier. A second pair of outputs from its sync-pulse section is taken to an *adder* stage for insertion into the video waveform, and the composite video signal is passed into the transmission cable.

On arrival at the monitor, the signal is first amplified, then passed to the cathode of the picture tube. A second path is to the *sync separator* stage which works on an amplitude-discriminating basis to strip off the sync pulses for application to the timebase generators. They work in just the same fashion as those in the camera to generate sawtooth currents in the scanning coils with which to deflect the scanning beam, but this time in the display tube. Thus we have the two electron beams – one at the sender and one at the receiver – swinging to and fro and up and down in perfect synchronism; and a flow of constantly-changing voltage in the form of the video signal conveying the pattern of light and shade from the vidicon target to the picture-tube screen.

The picture tube

In Chapter 7 we shall study the operation of colour tubes, and as an introduction to these, we need to examine the workings of monochrome tubes – they have much in common! Figure 1.10 shows the basics of a picture tube, which shares many features with the vidicon already described.

The process starts with a heated cathode, from which electrons are 'boiled off' by thermal agitation, to form a *space charge* around the cathode. Depending on the negative potential we choose for the 'grid' (in practice a cylinder surrounding the cathode, with a tiny hole in its otherwise-closed outer face), some of these electrons are attracted away down the tube neck towards the screen by the highly-positive anode cylinders, some of which are shaped and arranged to form an electron-lens whose focal point is the inner surface of the display screen. As the electron beam passes into the 'bowl' of the picture tube it comes under the influence of the scanning coils which deflect the scanning spot, at line and field rate, to trace out a rectangle of light, known as a *raster*, on the inner face of the tube. The screen is *aluminised*, and the thin aluminium layer is held at a potential

Figure 1.10 *The working principle of a monochrome picture tube*

9

of several thousand volts with respect to the cathode – this gives an accelerating 'boost' to the electrons in the beam, so that they collide with the phosphor screen surface at very high velocity indeed.

How is the light produced? The inner surface of the faceplate is coated with a continuous layer of *phosphor* which has the property of emitting white light when a high-velocity electron beam impinges on it. The aluminised screen backing reflects the light forward, and forms a barrier to prevent harmful ions burning the screen. The light output is regulated by varying the beam current, and as we have seen, this is the primary function of the 'grid' cylinder. We can arrange a fixed bias on this electrode to set the overall raster brightness, then feed a video signal to it (normally about 80 V peak-to-peak is required) to instantaneously vary the beam current to trace out, and faithfully copy, the brightness of each individual pixel as it is positioned on the screen by the scanning system. In practice the grid is held at a fixed potential and the video signal is applied to the cathode.

Bandwidth

The scheme outlined so far describes the stages in capturing, relaying and reproducing the picture. We have given little thought as yet to the requirements of the transmission medium, be it a cable, space or a glass fibre!

Mention has already been made of pixels, and the more we have, the faster the rate of change of the video facsimile signal. Much depends on the scanning rate, and this has to be faster than a certain minimum to avoid a disturbing flicker effect in the picture. By using the interlace technique we achieve 50 'flashes' per second, and this is just sufficient to accommodate the human eye's 'time-constant', known as *persistence of vision*. To capture reasonable detail we need all of 625 lines (in fact only 575 of them are used for the picture itself) and this means that if the 450 000 or so picture elements are each different from their neighbours, the rate of change of the video signal needs to be in the region of 5.5 MHz (MHz, megahertz, one million cycles per second).

Let us look more closely into the reasons for this. If we are televising a picture of a black cat on a white carpet, the rate of change of the video signal will be very low except at the point of the scanning spot's transitions from carpet to cat and back again, each of which will give rise to a sudden *transient*. At the other extreme, if the camera is looking at a pattern of fine vertical lines the video signal will be much 'busier' and contain a great deal of HF energy. In practice the frequencies in the video signal are mostly related to line and field scanning rates, and much of the energy in a video signal is concentrated into 'packets' centred on multiples of line and field frequency. This is an important point, and one to which we shall return.

Modulation

The word modulation, in our context, means the impressing of a signal waveform (usually video, sound or digital pulses) onto *a carrier* wave. The nature of the carrier wave is dependent on the medium to be traversed – space, cable, optical fibre; the distance required to be covered; and to a lesser extent, the nature of the information to be carried. Thus the medium-wave sound broadcast band (MF, medium frequency) is suitable for long-distance broadcasts of rather indifferent-quality sound signals, but quite useless for television; at the other extreme, the SHF (super-high frequency) band is well-suited to a 'beamed' television broadcast service from an orbiting satellite, but one wouldn't expect to find Radio Brighton up there as well! So far as the studio and viewer are concerned, the carrier is irrelevant, because it acts purely as a vehicle on which the wanted signal travels, being discarded at the receiving end once its usefulness has been realised.

Types of modulation

The four types of modulation used in communications to attach information to a carrier wave are (a) amplitude modulation, (b) frequency modulation, (c) phase modulation (which is a variant of *b*), and (d) pulse-code modulation. In ordinary sound broadcasting (a) or (b) is normally adopted, the former at LF, MF and HF, and the latter at VHF (very high frequency, around 100 MHz). In *terrestrial* TV broadcasts, (a) is used for vision and (b) for sound. In fact, television uses all four modulation modes, because a form of phase modulation is used for the colouring signal, as we shall see in later chapters, and modern TV links also make use of PCM.

Amplitude modulation

AM is amongst the earliest forms of modulation to be used, and probably the easiest to understand. It is perhaps surprising that this was pre-dated, however, by PCM (*d* above) in that the very first crude spark transmitters used this mode, conveying the message by simply interrupting the transmitter's operation by means of a morse key! We use AM for the vision signal in terrestrial TV broadcasts; let's see how it is arranged.

For AM, we start with a stable oscillator to generate the carrier wave. For TV broadcasts, the crystal oscillator runs at a sub-multiple of the station frequency, and its output is multiplied up to the required frequency (corresponding to the specified channel number). The information to be sent is made to vary the *amplitude* of the carrier wave, so that in our example of a TV broadcast, the waveform of Figure 1.8 is fed to a *modulator* which may control the *gain* of an RF amplifier which is handling the carrier wave. Thus the video signal is impressed onto the carrier, and the polarity of the video signal is arranged to give *negative modulation*. This means that the tips of the sync pulses give rise to maximum carrier power, and whites in the vision modulating signal raise only about one-fifth of that level. This is illustrated in Figure 1.11 which represents the waveform fed to the UHF broadcasting aerial, and present in the tuner and IF sections of a conventional TV receiver.

11

Figure 1.11 *AM modulation of the video signal onto an RF carrier for transmission*

One of the advantages of AM is the simplicity of the detection or *demodulation* process, which in its original form requires only a rectifier and filter to recover the modulating information.

Frequency modulation

This is the second most popular mode for broadcast use, and is used for high-fidelity (hi-fi) sound transmissions in VHF Band II. Most of these radio transmissions are multiplex-stereo-encoded. So as we are concerned here, the most significant use of FM modulation is for the TV sound transmissions which accompany 625-line UHF picture broadcasts, and for vision broadcasts from satellites. In this type of modulation the carrier amplitude is kept constant and the frequency of the wave is varied at a *rate* dependent on the frequency of the modulating signal and by an *amount* (deviation) dependent on the strength of the modulating signal (see Figure 1.12). Thus the 'louder' the signal the greater the deviation, while the higher its frequency, the greater the rate at which the carrier frequency is varied either side of its nominal frequency.

Maximum deviation of the TV sound carrier is about ±50 kHz (kHz, kilohertz, one thousand cycles per second) corresponding to maximum modulation depth of this particular system, commonly described as 100 per cent modulation. 100 per cent amplitude modulation is when the troughs of the modulated waveform fall to the zero datum line. This is the absolute physical limit for AM.

Phase modulation

Here the phase of the carrier wave is altered according to the modulating signal. It has much in common with frequency modulation, but the deviation is very small

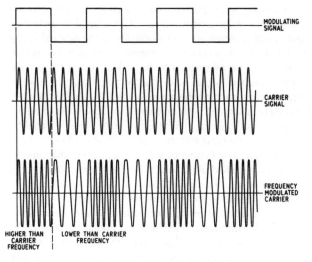

Figure 1.12 *Illustrating frequency modulation*

compared with the carrier frequency itself. In a television system, a common form of phase modulation is the transmission of a *reference* signal of constant period alongside a second carrier at the same frequency, but with phase, or timing, variations to convey the required information. This is the basis of the colour encoding system, fully dealt with in later chapters.

Pulse-code modulation (PCM)

An alternative method of modulation is a system of sampling, or quantising, an analogue signal to derive a transmission signal which has only two states, on and off. This is achieved by measuring the instantaneous amplitude of the sound or vision waveform at regular time intervals, short enough to ensure that the highest required frequency is adequately sampled. The amplitude of the signal at each 'sampling spot' is measured and assigned a number which represents the amplitude of the signal at that moment in time. This number can be represented in *binary* (two-state) form, as a series of pulses each of which has a closely defined time slot in a serial transmission. The presence or absence of a pulse in each time slot forms a pattern, unique to each number we may wish to send. The numbers are produced in a continuous string, as it were (known as *serial* transmission). In this form the signal is dispatched, a primary requirement of the transmission medium being sufficient bandwidth to enable it to resolve two recognisable levels (presence of pulse or absence of pulse) at high rates of change. Provided that this requirement is met, the sorts of distortion encountered by the signal on its path do not upset the final reproduction. The second requirement is that some form of *framing* or timing signal accompanies the digital transmission to synchronise the *decoder* at the receiving end.

13

On receipt, the PCM binary signal passes into a *decoder* which recreates 'numbers' from the pulse train. After processing in a D/A (digital to analogue) converter, the analogue signal is recreated, and its fidelity to the original waveform at the sending end depends on two factors – sampling rate and the number of quantising levels. The sampling rate needs to be at least twice the highest frequency of interest in the *baseband* signal (known as *Nyquist* rate). The number of quantising levels depends very much on the nature of the signal being sampled. For TV an impeccable picture is secured with no more than 256 levels of brightness. Surprisingly, perhaps, an audio signal, if it is to have an acceptable signal-to-noise (S/N) ratio and be free of spurious effects, requires in the region of 1024 quantising levels, though the sampling frequency will be much lower than that for TV pictures. As those who have studied digital electronics will know, this implies an eight-bit 'word' for each pixel of a TV signal ($256 = 2^8$) and a ten-bit word for sound encoding ($1024 = 2^{10}$). Some sound systems use 14-bit words.

PCM has many advantages over other modulation systems where the transmission path is long, and subject to distortion, reflections and similar hazards for the signal. It is finding increasing applications in television technology, especially in the areas of studio equipment, video tape recording, broadcasters' inter-location links, fibre-optic transmissions and even domestic receiving equipment, though while the transmission system and display device remain in analogue form the benefits of digital processing, particularly in the receiver, are somewhat doubtful.

Sidebands

Whatever type of modulation is used, sidebands are produced. Treated mathematically, the modulated signal can be shown to become a 'group' consisting of a central frequency (the carrier wave) with a pair of symmetrical 'wings' consisting of a number of signals whose displacement from the carrier wave and amplitude depend on the excursions of the modulating signal. This results in a spread of signals across the frequency spectrum, shown for AM modulation in Figure 1.13.

These are the sidebands, and the maximum spread is determined by the highest frequency to be transmitted as modulation. Thus a high-frequency modulating signal will produce sidebands which are located at the limits of the transmission bandwidth. In a vision broadcast these high-frequency modulating signals correspond to the fine detail in the picture, and in a sound broadcast to the upper-treble notes.

Figure 1.13 shows the central carrier frequency with sidebands for four distinct modulating frequencies simultaneously present. The outer sidebands A and A' would correspond to the high frequency (say around 5 MHz) components of the TV picture signal while D/D' would arise from lower-definition parts of the baseband signal. C/C' and B/B' represent intermediate modulating frequencies. Also see page 33.

In FM modulation systems sidebands are also present; their nature is somewhat different, however. Even if only a single pure tone modulating frequency is present, a series of sidebands is produced, and in theory, the 'spread' of the sidebands is infinite. They are spaced on either side of the carrier by multiples of the mod-

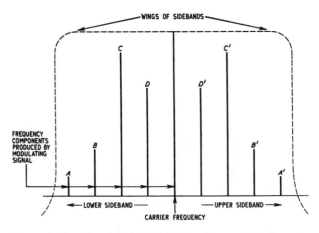

Figure 1.13 *Showing the sidebands of a modulated carrier wave*

ulating frequency, although only those relatively close to the carrier are significant in the reception process, depending on ultimate quality. In general an FM transmission system needs more bandwidth than an AM one, as demonstrated by the radio broadcast bands, where a receiving 'window' 20 kHz wide (though it is often less than this in practice, to minimise adjacent channel interference) would afford most adequate reception of MF AM broadcasts, whereas VHF FM reception demands a 'window' of 200 kHz (wider for stereo). For picture transmissions, the comparison is similar – terrestrial AM TV transmitters operate within a bandwidth of 8 MHz, while FM picture signals from satellites demand 27 MHz of spectrum space, although in both cases cited the FM mode affords more 'detail' in the modulating signal, and the bands used for the FM examples are higher in frequency, giving more 'elbow room', so to speak, for the greater sideband spread. We shall return to the subject of sidebands several times in later chapters.

Transmission bandwidth

This band of frequencies has to pass through the various types of equipment, including amplifiers, aerial systems, detectors and so forth in its passage from the studio or programme source to the viewer's home. The bandwidth of such devices will determine just how much of the detail of the original signal gets through. If the bandwidth is too small it will eliminate some of the 'detail' information and also, perhaps, distort the signal in other ways. If, on the other hand, the bandwidth were too great it would allow undue 'noise' and other unwanted signals to enter the system and thus detract from the quality of the vision (and sound). In any case the bandwidth spread of any transmission must be limited to the minimum possible spectrum space (rigidly-kept channel widths are specified for all types of transmission) to conserve valuable band space and prevent mutual interference with adjacent channels. A descriptive 'model' illustrating these points is given in Figure 1.14.

TRANSMITTER

f_o

GRADUALLY DIMINISHING SIGNAL

PHALANXES OF NOISE WHICH CAN BE PARTIALLY EXCLUDED BY REDUCED RECEIVER BANDWIDTH

TOO WIDE A BANDWIDTH RECEIVER ALLOWS IN TOO MUCH NOISE

TOO NARROW A BANDWIDTH RECEIVER CUTS DOWN NOISE BUT EXCLUDES SOME OF THE SIDEBANDS

Figure 1.14 *Model showing the relationship between the sidebands and the bandwidth*

2 Light and colour

We know radio and TV broadcasts as electromagnetic waves whose frequency determines their wavelength, and various broadcast bands have already been mentioned in Chapter 1. As we go up in frequency we pass through the bands allotted to radio transmissions, then terrestrial TV broadcast and space communications. Way beyond these we come into an area where electromagnetic radiation is manifest as heat, and continuing upwards we find infra-red radiation, and then a narrow band (between 380×10^6 and 790×10^6 MHz) which represents light energy. Beyond the 'light band' we pass into a region of ultra-violet radiation, X-rays and then cosmic rays.

Figure 2.1 gives an impression of the energy distribution curves for red, green and blue lights. It can be seen that they come in the same order as in a rainbow or from a prism. When we see white light it is in fact a mixture of coloured lights of all hues, and the 'splitting' effect of a prism demonstrates this by providing a different refractive index for each light wavelength. If the resulting colour spectrum is passed into a second prism it will be recombined into white light again!

This splitting and recombining process indicates that white light contains all the visible colours, and that by *adding* suitable proportions of coloured light we can create white light. Many combinations of colours may be added to render white, but in television three primary light colours are used: red, green and blue. The

Figure 2.1 *Energy distribution curves for red, green and blue light*

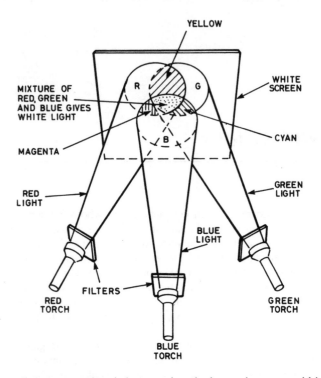

Figure 2.2 *White light is produced when red, green and blue lights are caused to overlap on a white screen. The nature of the white light depends on the intensities of the red, green and blue lights. Equal-energy white light is produced when the red, green and blue lights are of equal energy. This is a hypothetical 'white' used in colour studies. White light for colour television is called Illuminant D, which simulates direct sunlight and north-sky light mixture*

principle of *additive* light mixing, as it is called, is shown in Figure 2.2, where light of these three colours is projected onto a screen; in the centre of the display all three colours are present and they add to give a white light.

Colour filtering

Figure 2.2 shows each torch as having a coloured filter over its faceplate. It's important to remember that a filter absorbs all colours except that required, so that the red filter, for instance, will offer little resistance to the low-frequency end of the visible light spectrum, but attenuate green, blue and all other colours. Similarly, a green filter absorbs red, blue etc., allowing only the green component of the white torch beam to pass.

Colour temperature

It is difficult to define just what 'white' light is! The appearance of the white depends entirely on the strength and wavelength of each of the component primaries. Many monochrome picture tubes glow with a bluish, rather cold white, the result of a predominance of high frequencies in the rendered light spectrum; this is because that type of phosphor is more efficient in terms of light output.

Fortunately, an exact definition of the 'whiteness' of a light is available in the form of a *colour-temperature* which relates the nature of the white light to an absolute thermal temperature, that to which a black body must be raised to render the same 'colour' of white. For TV applications this has for many years been standardised at 6500 K, known as Iluminant D, and simulating 'standard daylight.'

Complementary colours

White light may be regarded as a mixture of red, green and blue lights. With these three at our disposal, other colours can be generated by various combinations of two. By removing the blue light we would leave a mixture of red and green lights, which would render yellow, as in Figure 2.2. Yellow is a complementary colour. It is, in fact, complementary to blue since blue was the additive primary which had to be removed from white light to produce it. By similar tokens the complementaries of red and green are cyan and magenta, which means that cyan (akin to turquoise) is produced by the addition of green and blue, and magenta (violet/purple) by the addition of red and blue.

Thus we have the three primaries red, green and blue, and the complementaries cyan, magenta and yellow; all these colours are obtainable – with white light – from the three primary colour *lights*.

It is difficult to visualise the wide range of hues that can be obtained from the three television primaries by changing their relative intensities; but those who view a good-quality colour television picture under correct (low!) ambient lighting conditions will appreciate that almost the full range of natural colours can be reproduced. It is noteworthy that in all the discussions and proposals for TV system improvement and picture enhancement, no change to the primary-colour *additive* mixing scheme has been suggested.

The chromaticity diagram

The range of colours can be conveniently represented by a chromaticity diagram, pictured in Figure 2.3. This is in fact a more elaborate extension of Figure 2.2, showing an elliptical area of various 'whites' at the centre, with the wavelengths of the various colours shown around the periphery. The colours between red and blue have no wavelength references, being 'non-spectral' colours resulting from a mix of components from opposite ends of the visible light spectrum.

What we cannot show in a line drawing is the fact that the colours gradually merge into one another around the edges of the diagram. The three TV primaries

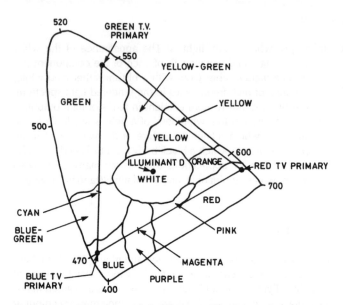

Figure 2.3 *The chromaticity diagram. The inner triangle rests on the TV primaries and the area enclosed by the triangle represents the range of colours in a TV display*

are shown, and linked as a triangle – it can be seen that for any primary, its complement appears about half-way along the opposite side of the triangle.

Luminance, hue and saturation

Basic television, as we saw in Chapter 1, is concerned solely with brightness variations, and the video waveform conveys information which describes only the variations of light and shade in the picture, as does a black-and-white photograph. As such it works only at a point in the middle of our chromaticity diagram, and so far as the viewer is concerned, the precise point depends on the makeup of the phosphor used in his picture tube! The monochrome signal with which we have been dealing so far is called a *luminance* signal.

To describe a colour, two more characteristics must be defined, and these are hue (colour) and saturation ('strength' of colour). Let's look at each in turn. The hue is set by the dominant wavelength of the light radiation, be it red, blue, cyan or whatever, and corresponds to points around the edges of our chromaticity diagram. Figure 2.3, however, encloses an *area* and real colours can occupy any point within that area. Very few colours in nature are fully *saturated* (which would set them on the periphery of the diagram); most fall within it, and this is where the second parameter, saturation, becomes relevant.

Saturation describes the amount by which the colour is diluted by white light, which in terms of the colour triangle in Figure 2.3, tells us how far from the white centre is the colour in question. If one can imagine a pointer whose origin is Illuminant D in the centre, and whose head pointed to the colour of interest, its length would represent saturation, and its *angle* would describe hue. Comparing the healthy pink of a child's face and the bright red of a London 'bus, it's not hard to see that both have the same hue – red – but the saturation levels are very different, the colour coming from the child's face corresponding to a point near the centre of the chromaticity diagram. This can be graphically illustrated on some colour TV receivers by lowering brightness and contrast and grossly over-advancing the colour level – the face will be reproduced in as strong a red as the 'bus in a normal picture!

The human eye

The phenomenon of persistence of vision has already been touched upon. The eye tends to retain an image for around 80 milliseconds after it has disappeared, and advantage is taken of this in both television and cinematography, where a serious of still frames create the illusion of a continuous moving picture. Other important characteristics of the eye are its relative insensitivity to *coloured* detail in a scene, and its failure to respond equally to all colours. Its greatest sensitivity is in the region of yellow/green, with lesser response in the areas of red and particularly blue.

Contrast ratio

In nature the range of brightness levels is infinite, from the brilliance of the sun to the total darkness of an enclosed cave. When an image is reproduced in photographic or TV form, the difference between the brightest and darkest parts of the picture is greatly reduced, and so far as TV reproduction is concerned, much depends on the level of ambient light. The darkest parts of the picture can be no blacker than the unenergised phosphor screen, and even with no external lighting source to illuminate the screen, a degree of reflection of the highlight picture components is present from the viewing area, and indeed the viewers' faces! Because the maximum brightness from a picture tube is limited, a contrast ratio of about 50:1 is usual in normal conditions. With a large light-absorbing auditorium and much higher light energy available, the cinema does much better than this.

Gamma correction

In the previous chapter we briefly examined the camera tube and picture tube in their roles of pickup and display devices. In terms of light input to voltage output, the vidicon tube is quite linear so that equal increments of brightness in the televised scene will give rise to equal steps of voltage at the target output point. Unfortunately, the picture tube is not linear in its operation. If we apply an equal-increment staircase waveform to the electrical input of the picture tube, the light output will not go up in corresponding steps; at high brightness levels the graduations will be emphasised or stretched, whereas low-key steps will be compressed. This means that the video signal needs to pass through a compensating circuit, with a gain/level characteristic equal and opposite to that of the tube. It would be expensive to provide such a gamma-correcting amplifier in every receiver, so the process is carried out at the transmitting end, to 'pre-distort' the luminance signal and cancel out the display-tube non-linearity. Figure 24.2 shows how the gamma correction system works at the transmitting end.

3 Reading and writing in three colours

If we bring together the themes of the last two chapters we are well on the way to realising a form of colour television system. Because all the colours in the scene to be televised can be analysed in terms of the three television primaries, we can assemble a colour TV outfit by triplicating the 'basic' television system and assigning one primary colour to each of the three. Figure 3.1 shows the set-up. Three identical cameras are used, each with an appropriate filter in front of its pickup tube. The three transmission cables carry video signals which correspond to the three primaries: R, G, and B for short. The monitors also have colour filters fitted, and their pictures are superimposed either by projection onto a common screen, or by a series of dichroic mirrors.

The colour picture reproduced by this set-up would be very good indeed once perfect superimposition, or *register*, of the three images had been achieved at both ends. As a system, however, its shortcomings are immediately obvious: three cameras, three separate transmission channels using (taking normal 625/50 parameters) a total of 16 MHz bandwidth, three identical receivers and goodness

Figure 3.1 *Elementary colour TV system working on the 'simultaneous' principle*

knows what in the form of filters and optics at each end! Not only this, but the system is quite incompatible with any monochrome (black-and-white mode) apparatus at either end.

The sequential approach

We know that basic television is no more than an advancing series of still frames, following each other quickly enough to prevent visible flicker. An early idea was to extend this technique to colour reproduction by mechanically changing the colour filter over the pickup tube at field or frame rate in the sequence of RGBRGB . . . so that 'snapshots' of each of the primary-colour components of the televised scene are sent out in quick succession. The single transmission system would now require only one TV receiver feeding a single picture tube equipped with three sequentially-switched colour filters working in synchronism with those at the camera.

The problem here is one of flicker. Because the repetition rate of each primary colour is now only one-third of the (just adequate) TV field repetition frequency, coloured objects flicker alarmingly; even if a form of 'storage' were devised to overcome this, moving coloured objects would traverse the screen in a blur or a series of hops. To overcome the problem a threefold increase in field-scanning rate would be required, calling in turn for three times the signal bandwidth throughout the system. Plainly, a radically different method of sending colour pictures was required, and the fact that much of the information in the R, G and B signals is common to all three (oscilloscope examination of these wave-forms on a normal picture defies anyone to tell them apart!) is a key factor in the solution devised.

Compatible colour television

Before we explore the practicalities of the colour TV system as it exists today, it is useful to provide a brief overview of how the problems described above are overcome in a modern system. In place of the three cameras we have a single camera which may contain between one and four pickup devices, with the necessary colour separation carried out by optical filters near or within the pickup tubes or sensors. The output from the camera is contained in one video transmission channel no more than 5.5 MHz wide by an encoding process, designed to render a signal which is compatible with monochrome receivers.

At the receiving end the encoded CVBS (colour, video, blanking and syncs) signal is handled in a single receiving channel identical to that of a monochrome set, so that only one tuner and IF amplifier are needed. The *luminance* and *chrominance* components of the signal are reassembled into RGB signals in a *decoder* and presented to a colour display device (usually a *shadowmask* picture tube) which is capable of reproducing all three primary-colour images on a single screen. This is summed up in Figure 3.2, and much of the first half of this book will be devoted to explaining the processes outlined above!

Figure 3.2 *Elementary concept of a complete colour television system*

Three-tube camera

Probably the simplest type of camera to understand is the three-sensor type, so we shall adopt this as a basis for an account of the operation of the first link in the chain. A multi-sensor camera has only one lens, of course, so the incoming light has not only to be colour-filtered, but 'split' and distributed between the three image sensors. This is achieved by a system of dichroic filters and silvered mirrors as outlined in Figure 3.3. The dichroic filter is a precision-engineered optical device capable of reflecting one colour while appearing transparent to others. In this way each pickup tube is presented with an image on its active surface composed of one primary-colour component of the picture to be televised.

The output from each image sensor is thus the electrical fascimile of the amount of each primary colour present in the scene. These outputs are now ready for processing to produce the compatible CVBS signal referred to earlier.

Figure 3.3 *Basic elements of the front end of a colour-transmitting system*

Deriving a luminance signal

It is essential that any colour system produces a signal which is recognisable to a monochrome receiver or monitor, so that the viewer has the choice between a colour or black-and-white set. This requires that the colour transmission conforms to all the specifications of the 'basic' TV system, so we must send primarily a luminance signal to convey brightness information, then add to it as best we can further information streams, used only in colour receivers, to describe the colour. This meets the compatibility requirement by permitting the use of monochrome picture sources at the sending end, and the choice of receiver at the receiving end. It has the incidental advantage of providing black-and-white reproduction when the receiver's decoder breaks down! Our first task, then, is to derive a luminance signal from the RGB outputs of the three pickup sensors.

The luminance signal is produced by adding the red, green and blue signals from the image sensors in the respective proportions of 30, 59 and 11 per cent, making a total of 100 per cent. This is best seen in simple algebraic terms such that the luminance signal, denoted Y in colour television, is $Y = 0.3R + 0.59G + 0.11B$. Here R, G and B correspond to the primary colours of red, green and blue. If each primary-colour signal from the corresponding image sensor is initially adjusted on a pure peak white scene for 1V, then by taking 30 per cent of the signal from the red sensor, 59 per cent of the signal from the green sensor and 11 per cent of the signal from the blue sensor and adding them all together we would obtain a total of 1V Y signal. The image sensors, of course, are not themselves coloured, any more than the green channels, blue electron beams and red guns we shall discuss later in the book! The 'colour labels' so applied merely refer to the circuit or channel in which the component in question is operating.

It may be wondered why equal proportions of R, G and B are not combined to produce the luminance signal. In fact the proportions chosen correspond to the sensitivities to each colour of the human eye, discussed in Chapter 2, and this ensures that the luminance display (the only one available to a monochrome set) appears to the viewer to be *panchromatic*.

This luminance signal, then, corresponds closely with that termed the video signal in a monochrome system. In fact, from first principles a monochrome (single-sensor) camera adjusted to yield 1 V signal from a pure white 'object card' would produce (under the same lighting conditions) 0.3 V output from a saturated red card, 0.59 V from a saturated green card, and 0.11 V from a saturated blue card – the same proportions as given from the pickup sensors of a colour camera scanning a pure white card, the three adding to produce 'unity' white signal. In practice the idea works well, though the necessary gamma-correction process introduces mild brightness errors on monochrome receivers in areas of the picture where heavily-saturated colours are present.

Single-tube cameras

In response to a healthy demand from users of portable VTR equipment, technology was developed for the practical realisation of a single-tube colour pickup

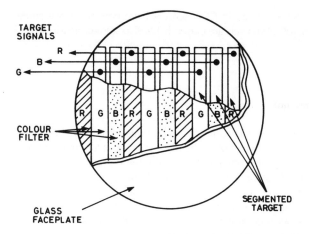

Figure 3.4 *Tri-electrode colour pickup tube by Hitachi – operation is explained in the text*

device. The single-tube camera has a lot to offer in terms of economy and simplicity since it obviates the need for expensive optical hardware, and poses no registration problems. These advantages have a trade-off in the realm of picture quality, however, but this is less significant when (as is usually the case) such cameras are used in conjunction with a domestic-format VTR machine whose inherent shortcomings mask those of the camera.

All single-tube cameras to date depend for their operation on a special target system, in which a matrix is present to give a colour-discriminating property. The matrix is in the form of an optical filter, and some tubes also have a matrix target. A good example of the latter type is the tri-electrode tube developed by Hitachi, and shown in Figure 3.4. Behind the glass faceplate of what is basically an ordinary vidicon tube is a striped optical filter consisting of vertical strips in the order RGBRGB etc. Behind this are three grilles, one for each primary colour; their vertical bars are physically aligned with the corresponding optical colour stripes. The grilles are in fact segmented targets which are scanned in the normal way by the vidicon beam, and the outputs rendered by these three stripe targets are RGB video signals corresponding to the outputs from the separate tubes in the conventional camera. A Y signal is obtained by adding correct proportions of R, G and B as before.

CCD colour cameras

The CCD pickup device described in Chapter 1 is used in colour cameras. It has the advantages over thermionic devices of small size, and physical and optical ruggedness. As with tubes, either three, two or one device can be used for full colour pickup, though single- and two-sensor cameras need some form of filter matrix along the lines of the stripe-filter vidicons. The early problems of low

27

definition, halation and 'vignetting' in CCD sensors have now been overcome, and devices with up to 500 000 pixels are currently fitted to semi-professional and 'home' cameras.

Colour-difference signals

We are now aware that a coloured scene possesses three important characteristics. One is the brightness of any part of it, already discussed as luminance; two is the hue of any part, which is the actual colour; and three is the saturation of the colour, that is the depth of colour. Any hue can vary between very pale and very deep, and the amount of the colour is basically its saturation. For example, a laurel leaf represents highly saturated green, while the pastel green of my blotting paper is the same hue but less saturated, which means that on a TV screen its image would have a greater contribution of white.

When an image sensor is scanning the red in a scene, as an example, it receives information on the luminance, hue and saturation because all three are obviously present in any colour. However, remember that the luminance signal is processed separately by proportioned addition of the primary-colour signals and transmitted effectively in 'isolation' so that it can be used by monochrome receivers. This, then, means that an additional signal has to be added to and transmitted with the luminance signal so that colour receivers will obtain the extra information they require to describe the hue and saturation of the individual pixels. This auxiliary signal may be regarded as the 'colouring' agent; it is called the *chroma* (short for chrominance) signal.

It is formed initially on a *subcarrier* which is then suppressed at the transmitter and recreated in the receiver, as we shall see later. The subcarrier is modulated in a special way by *colour-difference* signals, of which there are three but only two of them need to be transmitted.

The three colour-difference signals are red minus the luminance signal, green minus the luminance signal, and blue minus the luminance signal. By red, green and blue is meant the primary-colour signals delivered by the colour camera, while the luminance signal is the Y component as defined by the expression given earlier.

Thus, in simple algebraic terms the three colour-difference signals are: R − Y, G − Y and B − Y.

It is not necessary to transmit all three colour-difference signals. If we send Y, R − Y and B − Y, it is easy to recreate G − Y at the receiver. Let's see how this comes about. At the receiver we have available the Y signal, and information on how *different* is the R signal (in R − Y) and how *different* is the B signal (in B − Y). When these *combined* differences are taken into account, any outstanding balance (or debit!) represents the G difference, so that a G − Y signal can be derived by adding suitable proportions of *negative* R − Y and B − Y signals in the receiver's decoder.

The reason behind the choice of R − Y and B − Y signals for transmission is very simple. Although any two difference signals could be sent, we have seen that the Y signal draws more heavily on the green primary-colour component of the televised image than either of the other two (Y contains 0.59G), so it follows that the difference between G and Y is less than that for the others. This relatively small G – Y signal would be more vulnerable to noise in the transmission system than the larger R – Y and B – Y signals.

It is interesting to observe that the colour-difference signals fall to zero amplitude when the televised scene is devoid of colour - that is, when greys and whites are being transmitted. This is not really surprising when we remember that the chroma signal is composed of colour-difference components – in a *monochrome* transmission there is no difference between the primary-colour signals and the luminance signal! Monochrome programmes are becoming rare now, consisting mainly of archive material and old feature films.

We have seen that the red, green and blue signals from the image sensors of a colour television camera can be conveniently tailored to 'unity' (1 V) on a pure peak white input. Thus R = 1, G = 1, and B = 1. Y, we have seen, is equal to 0.3R + 0.59G + 0.11B, which means that on a pure peak white input we have 0.3(1) + 0.59(1) – 0.11(1), which equals 1. Clearly, then, from this we get R – Y = 1 – 1 = 0 and B – Y = 1 – 1 = 0. The same conditions exist on greys when the red, green and blue signals from the sensors are less than 'unity' but still equal. On colour scenes, of course, the RGB signals are not equal and so colour-difference signals arise, and only when this happens is a chroma signal produced.

It is possible, of course, to calculate both the luminance signal and the colour-difference signals from the colour scanned by the colour camera, remembering that anything below full saturation means that white is added to the predominant hue in terms of the three primary colours in the proportions required for the luminance signal. Thus, while the R, G and B signals become unequal when the camera is scanning a coloured scene, the Y signal still retains the proportions of 0.3, 0.59 and 0.11 of the R, G and B signals respectively.

For example, purple of below full saturation comprises a mixture of red and blue with a little green signal too, so that the Y proportions of the red, green and blue provide the 'white' which reduces the saturation. Thus we may have R = 0.6, G = 0.1 and B = 0.5, meaning that the luminance signal Y is equal to 0.3(0.6) + 0.59(0.1) + 0.11(0.5), or 0.18 + 0.059 + 0.055 = 0.294. Using this for Y, then, R – Y is 0.6 – 0.294 = 0.306 and B – Y 0.5 – 0.294 = 0.206. When the Y signal is of a greater voltage than the primary colour components of the colour-difference signal, the colour-difference signal as a whole then assumes a *negative* value as, of course, would be expected. The colour-difference signals, therefore, can swing from zero to a maximum in both the positive and negative directions. A little thought will confirm that if we have a white raster displayed, and wish to change it to blue or yellow, the means of doing so is to provide a positive or negative blue colour-difference signal. For saturated blue, the red and green colour-difference signals would also operate to turn off their respective colours.

Colour-difference signal bandwidth

The colour-difference signals are used, as we have seen, to *add* colour information to the basic black-and-white picture. Because the human eye cannot resolve fine detail in colour there is little point in transmitting high-definition colour-difference signals, which would be difficult to accommodate in the signal channel, and wasted on arrival! The luminance signal carries all the fine detail in the picture, then, and the colouring signal is 'overlaid' on the display in much coarser form. In practice the illusion is well-nigh perfect, and subjective viewing of the combination of high-definition luminance and rather 'woolly' chrominance is perfectly satisfactory. In the American colour system they go a step further, and transmit even lower definition in their 'Q' signal, corresponding to blue shades, where research shows that the eye is least able to discern detail. Reproduced colour pictures from domestic videocassette recorders offer even poorer chrominance definition, but still with (just) satisfactory results.

In the UK broadcast colour system, then, we limit the bandwidth of each colour-difference signal to about 1.2 MHz by means of electrical bandstop filters.

The encoder

The three primary-colour signals from the three image sensors are communicated to the input of the encoder, and this processes them ready for transmission. The process can be likened to the sending of an important message in code. If a code book is used at the dispatch end, the letter can be reduced to a shortened form and sent to the recipient who, using a similar code book, can decipher it and thereby recreate the original message.

The three primary-colour signals are first added together in the proportions required for the luminance or Y signal, as already explained. The resulting Y signal is then separately subtracted from the red and blue primary colour signals to give the two colour-difference signals R – Y and B – Y. These two signals are amplitude-modulated in quadrature (see anon) upon a subcarrier which is subsequently suppressed so that only the sidebands of the V and U signals remain.

In monochrome television (now largely confined to industrial surveillance and special-purpose applications) only the luminance signal and the sync and black level are modulated onto the broadcast carrier wave.

With the need to send a chroma signal in addition to the basic VBS (video, blanking and syncs) signal, coupled with the requirement to keep the combined signal within the channel bandwidth normally occupied by a monochrome signal, the method of fitting together all the components of a signal for compatible colour television is necessarily more complex. As already outlined, the scheme utilises a colour subcarrier of a much lower frequency than the main carrier wave. The latter, in fact, may be in the hundreds of MHz or the GHz (GHz, gigahertz, one thousand million cycles per second) range, whereas the former is a few MHz only. It is within the luminance bandwidth range in fact, the actual frequency being geared to the line and field timebase repetition frequencies.

We shall be seeing later that to get the chroma signal to integrate neatly with the luminance signal, the frequency of the subcarrier must be related to line- and field-scanning frequences. The extra 'colouring' information is then squeezed into discrete intervals of low energy between the luminance sidebands in the overall frequency spectrum. It is by this technique that the extra chroma information can be carried in an ordinary 625-line television channel with the least mutual interference, especially to black-and-white pictures produced by monochrome receivers working from a colour signal.

This clever technique, of course, calls for a comprehensive type of sync-pulse generator, for its master oscillator must be correctly related to the frequency of the subcarrier generator. However, in practice the master oscillator generates subcarrier frequency, which is then divided down by counter circuits to derive line and field synchronisation pulses. This ensures that the relationship between the three is correct. Figure 3.5 outlines the processes described thus far.

Figure 3.5 *Showing in basic terms how the composite signal is developed. The subcarrier generator provides a reference for both sync and colour signals, and the composite signal developed contains all information necessary for the recreation of a colour picture*

Composition of the chroma signal

Having discussed the derivation of luminance and colour-difference signals, and seen how all the characteristics of a full colour picture can be carried in them, it's time now to explore the way in which they are combined to make up a CVBS (chroma, video, blanking, syncs) signal. We have already referred to the process of quadrature modulation onto a subcarrier - just what does this mean?

The basics of amplitude modulation have been discussed in Chapter 1. The modulation signal of lowish frequency modulates the carrier wave of higher frequency by causing its amplitude to vary in sympathy with the modulating signal. Figure 3.6 illustrates the effect. Here a single sinewave signal is shown modulating a higher-frequency carrier wave. This modulation is quite easy to understand from an elementary viewpoint, and it can be achieved by the use of simple circuitry. For example, the modulating signal can be applied to a transistor so that it effectively alters the supply voltage on the collector. Thus when the carrier wave is applied to the base, the amplitude of the output signal alters to the pattern of the modulation waveform. When a pure sine wave is the modulation signal, therefore, the carrier-wave amplitude varies in a like pattern and the carrier

31

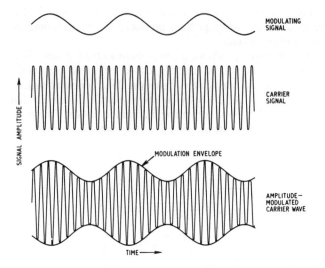

Figure 3.6 *The amplitude modulated waveform is produced by the carrier signal being modulated with the modulating signal. Note that the waveform inside the modulation envelope results from the addition of the carrier wave (fc), the upper sideband (fc + fm) and the lower sideband (fc − fm), where fm is the modulation frequency*

wave is then said to possess a modulation *envelope* as shown at the bottom of Figure 3.6.

We have already seen that any type of modulation gives rise to sideband signals, and for amplitude modulation there is an upper and a lower sideband for each modulation frequency. If the carrier wave is, say, 10 kHz and the modulating signal a pure 1 kHz sine wave (e.g. a single modulation frequency) then the upper sideband will be 10 + 1 = 11 kHz, while the lower sideband will be 10 − 1 = 9 kHz. This simple arithmetic follows for all single modulation frequencies. Thus the modulator delivers three signal components, the carrier wave, the upper sideband and the lower sideband. The information due to modulation is effectively present in the sidebands, so it is possible to suppress the carrier after modulation for transmission, though for demodulation at the receiver the carrier wave will need to be somehow reintroduced very accurately.

So much, then, for the amplitude modulation of one set of information, but what about the modulation of two sets of information which convey the R −Y and B − Y colour-difference signals? This is where the quadrature amplitude modulation of the subcarrier comes in. In the UK system of colour television the subcarrier frequency is accurately controlled at 4.43361875 MHz. This is usually referred to roughly as 4.43 MHz. The numerous decimal places (the last of which defines the frequency to one-hundredth of one cycle per second!) are necessary for various reasons, one of which is to minimise interference of 'dot-pattern' type which can mar received pictures due to a beat effect arising between subcarrier and line timebase frequencies. The rigid frequency relationship set up between these two gives optimum performance not only in minimising the dot pattern, but in preventing a disturbing 'crawl' effect of such pattern elements as remain. It is also necessary to have the subcarrier within the video bandwidth, and at as high a frequency as possible for minimum interference. In the PAL system both the upper and lower chrominance sidebands are exploited equally, and this means that subcarrier frequency must be chosen so that both sidebands, each extending to about 1.2 MHz, can be fully accommodated within the video spectrum, as shown in Figure 3.7.

Figure 3.7   *Spectrum of 625-line PAL signal in UK television channel*

The master sync-pulse generator, then, produces an output at subcarrier frequency, and this is fed into *two* amplifiers which produce two subcarrier outputs, but with one having a 90 degrees phase difference with respect to the other.

'Quadrature' and degrees of timing

A complete cycle of alternating current or voltage can be regarded as occupying 360 degrees – a circle! Although this statement is somewhat arbitrary it is supported by sound reasoning. Consider a generator, for example; this will, in the simplest case, rotate through 360 degrees and during that period produce a complete sine wave. If a second generator is set going but a quarter turn ahead of the first, then the sine wave produced by this will be 90 degrees out of phase with that yielded by the first one.

Since in the case of the colour television the one oscillator drives two amplifiers, the frequency of their two outputs will be absolutely identical and, moreover, the synchronism will be maintained, but always with one output 90° ahead of the other owing to the deliberately contrived 90° phase shift introduced by the circuit elements.

Figure 3.8 highlights the situation, where the two full-line sine waves are of exactly the same frequency but 90° apart in phase. The X axis is calibrated from 0 to 360° corresponding to a complete cycle of the first full-line sine wave. The second one starts a little later in time; in fact, when the first has arrived at the 90° mark. Thus we have a direct illustration of the 90° difference between the two

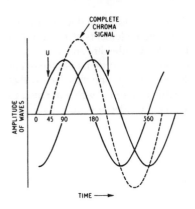

Figure 3.8 *The V and U waveforms have the same frequency and amplitude but differ in phase by 90°. The broken-line waveform represents the chroma signal complete which is the quadrature addition of the V and U signals*

subcarriers which, remember, are derived at the transmitter from a common oscillator or generator. Hence phase can be seen to be a function of *timing* between two signals. The term quadrature comes from the fact that there are four 90° segments in a circle; two signals with a 90° phase relationship are said to be in quadrature.

Colour modulation

One of our subcarrier signals is amplitude-modulated with the R –Y signal, then, and its quadrature companion is amplitude-modulated with the B – Y signal. After suitable *weighting* they become V and U chroma signals of the PAL colour system. Weighting is a process of reducing the amplitudes of the R – Y and B – Y subcarriers. At a later stage they will be added to the luminance signal, and where a large subcarrier signal (corresponding to a highly-saturated colour) coincides with an extreme excursion of the luminance signal (corresponding to a very light or very dark part of the picture) the combination of the two could lead to over-modulation of the transmitter. The weighting values are: V = 0.877 (R – Y) and U = 0.493 (B – Y). After recovery in the receiver's decoder, the subcarrier signals are restored to normal proportions by simple adjustment of amplifier gain in each colour-difference signal path.

The next move in the game is to add the V and U signals to form the chroma signal proper, and this signal is shown by the broken-line sine wave in Figure 3.8. The fundamental feature of this mode of signal addition is that by special detection at the receiving end it becomes possible to isolate the V and U signals again and thus extract the original R – Y and B – Y modulation signals. This is facilitated by the fixed 90° phase difference between the two original carriers. Signals of the same frequency and phase lose their individual identity for all time when added.

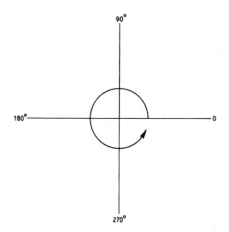

Figure 3.9 *Vector 'background' diagram in four quadrants. Time or motion is regarded by convention as being anticlockwise from the 0-degree datum, as shown by the arrowheaded circle*

In colour TV engineering, the various colouring signals are regarded in terms of vectors. These are mathematical devices for displaying the specific features of amplitude and phase of a signal at one instant in time. The basic vector 'background' is given in Figure 3.9 where our complete timing circle is represented by the four quadrants. Motion or time is regarded as anti-clockwise, so starting from the zero-degree datum we follow the angles of phase as shown by the arrowed circle. Colour signals are generally more complex than implied by simple vectors and phasors, so this method of presentation may be regarded as a simplification, though an adequate one for our purposes here.

Developing the vector theme in terms of the V and U signals in Figure 3.8 we get the vector diagram of Figure 3.10. The arrowed lines here correspond to the three signals – the V and U signals and the resulting chroma signal. The amplitude of this result is obtained by completing the square (or rectangle, when V and U

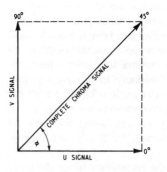

Figure 3.10 *Vector representation of the V, U and complete chroma signals of Figure 3.8 (see text)*

signals are not equal) as shown. The diagram clearly shows that the V and U signals have a 90° phase difference and that the amplitudes of these two signals are in this case equal. The angle changes (but within the same quadrant) when V and U amplitudes differ. Should the amplitudes alter together and in sympathy, then the original angle is maintained, though the vector line will now be shorter to describe lesser saturation in the transmitted colour. That the complete chroma signal is 45 degrees relative to either the U or V signal is proved in Figure 3.8. However, it is more convenient to work with vectors than with complex waveforms and the remainder of our account will be on a vector basis.

We can easily discover the amplitude of the complete chroma signal which, in colour television at least, is often referred to as the phasor, from the expression: phasor amplitude $= \sqrt{(V^2 + U^2)}$.

Thus when the V and U signals have equal amplitude, the phasor length is 1.4 times that of V or U. Readers with some knowledge of trigonometry will also see that the angle ϕ of the phasor in the quadrant of Figure 3.10 is equal to $\tan^{-1}V/U$.

We next need to get clear what effect negative colour-difference signals have on the vector diagram. The quadrant of Figure 3.10 can only accommodate reds, purples and blues, and we need to describe all the colours within the triangle in Figure 2.3. To enlarge on our earlier discussion of colour-difference signals, we saw that they can move from zero in a positive or negative direction. In terms of fully-saturated colours, let's take one or two examples to illustrate this. Refer to Figure 3.11, which shows a standard colour-bar display, consisting of a white bar, followed by saturated bars of yellow, cyan, green, magenta, red and blue, then a black (zero luminance and zero colour) bar, reading from left to right. On the first bar, all three lights are on to give a full-brightness equal mix of R, G and B. There is no colour present, so the colour-difference signals are at zero as shown in the time-related luminance and colour-difference waveforms below. Moving on to the yellow bar, we know that this consists of a mixture of red and green light, so the colour-difference signals for these two will have to act to turn them up, giving rise to positive colour-difference signals, while the blue light needs to be turned off, calling for a *negative* B – Y signal. The result is the complement of blue–yellow. For the next (cyan) bar, we are looking to turn off red, and turn up blue and green (to compensate for the loss of *brightness* output from the now-absent red light), so here we have positive B – Y and G – Y signals, along with a negative R – Y signal. The other bars call for different combinations of colour-difference signals, but their relationship is plain from the waveforms below. Relating these to the vector diagram, let us take the case of the yellow bar. Here the B – Y signal necessarily moves in a negative direction, into the second quadrant of the vector diagram (Figure 3.12). This indicates how the chroma subcarrier reflects the polarity of the colour-difference signals in terms of its phase. In the case of yellow, the R – Y signal is still positive, so the resultant vector angle reflects the presence of a small positive value of R – Y and a large negative value of B – Y. Figure 3.12 shows the vectors for all three complementary colours, along with that for green, which (as should now be plain) calls for negative R – Y and negative B – Y signals.

To summarise, however, Table 3.1 gives all the parameters so far discussed relative to the hues of the standard colour bars. All these values are based on 100 per cent saturation and amplitude.

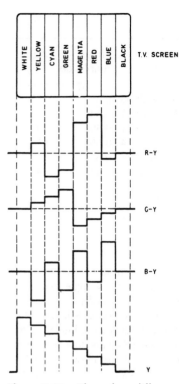

Figure 3.11 *The colour-difference signals and their relationship to the TV screen display on standard colour bars*

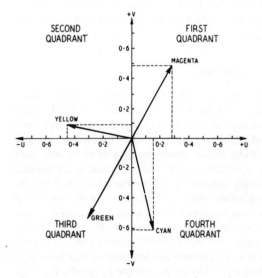

Figure 3.12 *Phasors for various colours: the vector angles and lengths for green and complementary colours megenta, yellow and cyan*

Table 3.1

Colour	Y	B − Y	R − Y	U	V	Phasor Amplitude	Angle (deg)
Yellow	0.89	−0.89	+0.11	−0.4388	+0.0965	0.44	167
Cyan	0.7	+0.3	−0.7	+0.1479	−0.6139	0.63	283
Green	0.59	−0.59	−0.59	−0.2909	−0.5174	0.59	241
Magenta	0.41	+0.59	+0.59	+0.2909	+0.5174	0.59	61
Red	0.3	−0.3	+0.7	−0.1479	−0.6139	0.63	103
Blue	0.11	+0.89	−0.11	+0.4388	−0.0965	0.44	347
White	1.0	0	0	0	0	0	–
Black	0	0	0	0	0	0	–

This brings us to the concept of a 'colour clock' whose face is the vector diagram, and which has a single 'hand' able to rotate to any angle to describe the hue of the picture and having a 'telescopic' feature, whereby the longer the hand the greater the saturation. With the hand pointing north, as it were, fully saturated red would imply a long hand, whereas pale pink would give rise to a short one. Similarly, a bright orange colour would bring the hand to a 'long north-easterly' aspect and a green colour would direct it to the south-west. It will be remembered that the same 'telescopic pointer' effect was used in our description of the chromaticity diagram, Figure 2.3. Thus as each line of picture is scanned, so can be visualised the phasor changing in both amplitude and angle to describe the saturation and the hue of each individual pixel in the line in turn. When there is no colour, the phasor shrinks to zero and the picture is then under the sole control of the luminance part of the signal.

Integrating luminance and chrominance

We have now seen how all the colouring information is contained in a single chroma signal, by means of (a) leaving out the G − Y component and (b) quadrature-modulating the remaining two colour-difference signals onto a single carefully-chosen subcarrier. The next step is to combine this with the luminance signal in a way which causes the minimum mutual interference – a daunting task, since the luminance signal appears to occupy all the available channel bandwidth! Close examination of the frequency spectrum of the luminance signal, however, reveals that most of the picture information is centred on 'energy packets' at multiples of line- and field-scanning frequencies, as intimated in Chapter 1. The spaces between these packets are relatively quiet. Let us now give some thought to the chrominance subcarrier signal. It is describing the same picture, and many 'detail' features of that picture will be common to chrominance and luminance signals. Thus the chroma sidebands will have a packet energy-distribution characteristic similar to that of the main signal, and this is the key to the interleaving process which is used. If we can *offset* the chroma subcarrier frequency to place it exactly between two of the luminance packets, as it were, the chroma sidebands will fall between the similarly-spaced luminance energy packets, rather like the teeth of two interlocking combs. Figure 3.13 illustrates the principle; to achieve it the subcarrier frequency is related to the line-scanning frequency thus:

$$\text{Line frequency} \quad (f_1) = \frac{f_{sc} - \frac{1}{2}f_f}{284 - \frac{1}{4}}$$

where f_{sc} is the subcarrier frequency (4.43361875 MHz for PAL), f_1 is the line frequency (15.625kHz) and f_f the field-scanning frequency (50Hz).

Figure 3.13 *The sidebands can be looked upon as packets of energy at various frequencies. This diagram shows the interleaving of chrominance and luminance, and the situation of the sound channel in the transmitted signal*

Suppression of subcarrier

Because the modulating information is wholly present in the sidebands of the subcarrier, it is possible to suppress the subcarrier itself at the transmitter, and by so doing we can significantly reduce the severity of the dot-pattern introduced at the receiver by the presence of the colour signal. It will be recalled that the waveform inside the modulation envelope such as that shown in Figure 3.6 results from the addition of the carrier wave and the upper and lower sidebands. As would be expected, a modulation waveform devoid of its carrier wave differs significantly from that with the carrier wave intact since the former is composed of only the upper and lower sidebands. Figure 3.14 attempts in one way to reveal the salient features.

Compared with the bottom diagram in Figure 3.6 it will be seen that the envelope has effectively 'collapsed', so that the top and bottom parts intertwine, the sine wave in heavy line representing the top of the original and that in light line representing the bottom of the original. Further, the high frequency signal inside

39

Figure 3.14 *Modulation waveform when the carrier wave is suppressed. Compare this with the waveform at the bottom of Figure 3.6. An important aspect is the phase reversal of the high-frequency wave each time the modulation sine waves cross the datum line. This waveform is the addition of the upper and lower sidebands only*

the collapsed envelope has also changed character. The frequency, however, is just the same as the original carrier wave because it is composed of the original upper and lower sidebands, the average of which is the carrier frequency; but it can be seen that phase reversals occur each time the sine waves representing the top and bottom parts of the original envelope pass through the datum line. It is difficult to show these diagramatically; but the high-frequency signal changes phase by 180 degrees at each 'envelope crossover' point, and this has a vital significance in colour encoding, as we shall see.

The subcarrier modulation constitutes the B – Y and R – Y signals, which change continuously during a programme. However, in a colour-bar signal the signal is less 'busy' because the colour signals remain constant over each bar per line scan, changing only from bar to bar.

Figure 3.15 shows the weighted B – Y signal produced by the yellow and cyan bars at (a), the U chroma signal modulation waveform due to the bars at (b) and the reference subcarrier signal at (c). The U chroma modulation signal at (b) is with the subcarrier suppressed, and since the modulation signal is a stepped waveform going from – 0.33 (corresponding to the yellow bar) to + 0.1 (corresponding to the cyan bar) through the zero datum, it follows that the polarity change between the two bars will cause a 180° phase reversal of the high-frequency signal just the same as when the modulation is a sine-wave signal, shown in Figure 3.14.

The phase reversal is indicated, and more clearly shown in Figure 3.15 than in Figure 3.14. For example, it will be seen that the positive tips of the subcarrier correspond to the negative tips of the U chroma signal during the yellow bar, and to the positive tips (owing to the phase reversal) during the cyan bar. Thus if we can recreate the original subcarrier at the colour receiver, and make it accurately sample the peaks of the chroma signal in this way, information is recovered on both the amplitude *and* the polarity of the colour-difference signal. This, in fact, is how the chroma signals are demodulated at the receiver. The operation of its subcarrier generator will be described in the next chapter.

Composite colour signal

When we add our carefully-contrived suppressed subcarrier signal to the luminance waveform we form the composite colour signal (CVBS signal) referred to

Figure 3.15 (a) PAL-weighted B – Y signal at the yellow/cyan bar transition. (b) The waveform resulting from suppressed carrier modulation. (c) Subcarrier signal in correct phase. At the receiver the reintroduced sub-carrier samples the peaks of the modulation sideband components to give an output corresponding to the amplitude and polarity of the original colour-difference signal. The polarity is given because the phase of the sideband components reverse each time the colour-difference modulation signal crosses the zero datum line (also see Figure 3.14)

near the beginning of this chapter. We shall have the basic television picture signal of our Figure 1.6 with the addition of the chroma signal, which latter will tend to ride on the luminance level, whatever that may be. Figure 3.16 indicates the form of the CVBS signal as it leaves the studio en route to the transmitting site. This is for a colour-bar signal, showing its luminance step waveform and the large superimposed subcarrier signal characteristic of fully-saturated bars. In a typical 'real' picture, saturation will not be nearly so high, and as a result the subcarrier amplitude will be correspondingly lower. The luminance signal, too, will be of a random nature, so that the equivalent composite waveform would appear fuzzy and gauze-like.

There is one other feature of Figure 3.16 which we have not yet considered, and that is the burst of what appears to be a chrominance signal on the back porch.

This is in fact a reference subcarrier timing signal for the receiving decoder, and it will be considered in detail in the next chapter.

Figure 3.16 *Oscillogram showing the composite signal over two lines of a colour-bar transmission*

The deciphering process at the receiver

Assuming that our CVBS signal passes unscathed through the transmission system, whatever form that may take, it will arrive at the receiver's detector output (or the monitor's vision input) in the same form, and now requires sorting out to derive the original RGB signals as were present at the output from the camera. The processing of the Y signal, as in a monochrome set, consists purely of amplification as it is already in a form recognisable to a picture tube. Our concern here is to examine the overall process of colour demodulation, and see how the chrominance signal is unscrambled and recombined with luminance to recreate R, G and B signals; and finally how they are displayed on the viewing screen.

The colour subcarrier is selected from the composite signal by means of a bandpass filter, on the left of Figure 3.17. This has a response about 2.4 MHz wide, centred on 4.43 MHz to embrace the entire sideband spread of the chrominance signal. We now need to demodulate the signal, and to achieve this the subcarrier is routed to two separate switches, each capable of operating at sub-

Figure 3.17 *A basic decoder, showing the synchronous detector-switch system and the derivation of a G – Y signal*

carrier rate. Imagine that the switches are normally off, and that each has a storage capacitor connected to its output. In this state the stores will be empty and nothing will happen until the switches close. In fact the switches are closed for a brief instant once every subcarrier cycle. Consider the upper switch S1 which is driven by a local source of subcarrier reference. If the phasing of the reference is correct, the switch will close momentarily at the instant when the V (carrying R – Y) carrier is at its peak, and the storage capacitor C1 will acquire a charge corresponding to the instantaneous level of the V signal during the *sampling* phase. At this time the U signal (carrying B – Y information) will be passing through zero because of the quadrature relationship between V and U subcarriers, so that the U signal cannot affect the charge acquired by C1. If this storage capacitor is paralleled by a resistor to give an appropriate time constant, the signal appearing across the resistor will accurately follow the V signal modulated onto the subcarrier at the sending end.

The U detector works in like fashion, but here the sampling phase must match that of the U signal. This is easily achieved by introducing a matching phase shift in the reference subcarrier path to the switch, as shown in Figure 3.17. S2, then, will switch on briefly at the instant that the U subcarrier is passing through its zenith and this of course represents the time of passage of the V subcarrier through the zero datum line. The signal appearing across R2 will be a replica of the U signal at the studio.

C1 and C2 can charge to negative levels as well as positive ones, and will do so whenever the narrow and precisely-timed 'sampling window' catches the sub-carrier below the zero datum line. It will be recalled that we discussed the ability of a subcarrier signal to reflect polarity changes in the colour-difference signals in terms of phase (Figure 3.15), and here we can see how the demodulators recover this information to present negative or positive signals at their outputs, dependent on the phase of the incoming subcarrier.

In practice we shall only see maximum positive outputs from the V and U demodulators on a saturated magenta picture; on the green bar, both will be at a maximum negative level, giving rise to a highly positive G – Y signal which, remember, is derived by adding *inverted* R – Y and B – Y signals. We shall come shortly to the G – Y recovery matrix. It is important that the triggering pulses for the switches, i.e. the local subcarrier signal, are precisely in phase with the broad-caster's subcarrier generator. If any timing error is present here, the sampling phases will take place at the wrong times relative to the incoming chroma signal, and this will cause incorrect or random colour-difference signal levels and pola-rities, making nonsense of the hues in the reproduced picture!

After detection and filtering, the V and U signals are de-weighted by passing them through amplifiers whose gain is the reciprocal of the weighting factor. Thus the V signal will undergo amplification by $1/0.877 = 1.140$ to render R – Y, and the U signal will be amplified by $1/0.493 = 2.028$ to render B – Y.

G – Y and RGB matrixing

Now that we have the signals back in R – Y and B – Y form, we can recover the missing G – Y signal. It will be recalled that 100 per cent Y signal is equal to $0.3R + 0.59G + 0.11B$ which, of course, is equal to $0.3Y + 0.11Y$. Now if we subtract the second expression from the first we get $0 = 0.30(R - Y) + 0.59(G - Y) + 0.11(B - Y)$, which gives $-0.59(G - Y) = 0.30(R - Y) + 0.11(B - Y)$ and from which is obtained $-G - Y = 0.30/0.59(R - Y) + 0.11/0.59(B - Y)$. Inverting both sides of this final equation renders: $G - Y = 0.3/0.59 - (R - Y) + 0.11/0.59 - (B - Y)$. Thus to arrive at a correct $G - Y$ signal we need to add 30 fifty-ninths of an *inverted* $R - Y$ signal to 11 fifty-ninths of an *inverted* $B - Y$ signal. This is shown in Figure 3.17, where R3 and R4 select the correct proportions.

Now we are back to the three separate colour-difference signals plus a Y signal and all is plain sailing! We merely add Y separately to each of the colour-difference signals to render RGB signals ready for the display device.

Display

In our earlier example three separate display devices were used, one for each primary colour. Except in some forms of projection display this is not common, and as is well known a colour picture tube is generally used, in which all three primary colour signals are handled simultaneously. We shall cover colour picture tubes in detail in Chapter 7; suffice it here to say that the shadowmask picture tube manages to simulate the effect of three superimposed tubes, each working in one primary colour. It has a single electron-gun assembly and is set up with accelerating voltage and deflection fields in just the same way as a monochrome tube, but its three cathodes accept the RGB signals with which we started this chapter. Thus we are able to read and write in three colours to capture, transmit and display a full colour picture, with but a single link between programme source and viewer.

There is more to the mechanics of encoding and decoding the colour signal, and indeed other ways of going about it. We shall explore these in the next chapter.

4 *The PAL system*

In the last chapter, the principle of chroma encoding was explained with particular reference to the chroma signal itself, and the method of 'dovetailing' the chroma and Y signals. Reference was made to the necessity of accurately reproducing the original subcarrier reference signal at the receiver so that chroma signal phase detection can be carried out with precise timing to faithfully reproduce the hue and saturation of the colour picture. To recreate the subcarrier at the receiving end, we must transmit some reference signal to regulate the 'clock' which generates *local* subcarrier to time the sampling phases of the V and U detectors.

The colour burst

Since the subcarrier is of constant and unvarying phase, we need not send a continuous reference signal. Provided we can arrange a very stable oscillator at the receiving end (invariably a high-Q crystal oscillator) we need only send a sample for comparison at regular intervals. A place has to be found for this sample in the CVBS waveform, and the only space left which can be utilised is the 5.8 µs back porch between the line sync pulse and the start of picture information. Onto this back porch is inserted ten cycles of subcarrier signal, with a peak-to-peak amplitude equal to the height of the sync pulse itself (300 mV in the standard-level CVBS signal). This *colour-burst* can be seen in Figure 3.16 and is the key, so far as the receiver's decoder is concerned, to the accurate regeneration of the subcarrier signal.

At the decoder, a *burst-gate* is present, triggered from the transmitted line sync pulse to open during the back porch. This isolates the burst signal and directs it into a *phase-lock-loop* (PLL) circuit which compares the frequency and phase of the local crystal oscillator with that of the bursts coming at 64 µs intervals: see Figure 4.1. Any discrepancy gives rise to an error signal from the phase detector; applied to a reactance stage associated with the crystal oscillator it can pull the oscillator into lock with the transmitted colour burst so that its frequency and phase are identical with that at the transmitting end. Figure 4.1, then, gives the bones of a complete 'basic' decoder system as an enlargement of Figure 3.17, and this basic method of encoding and decoding colour in a compatible system is that

Figure 4.1 *Driving the synchronous demodulators. The two transmitted components of chroma and burst are processed to render locked and correctly-proportioned R, G and B signals*

adopted in the USA, where it is known as the *NTSC* system (short for National Television Systems Committee, who recommended it for America in the early 1950s).

NTSC and PAL

The basic colour system outlined above works well, and has been in successful use in the USA for well over 40 years. Accurate decoding is facilitated by the fact that any distortions or timing errors encountered and picked up by the signal in the transmission system are imposed equally on the burst and chroma signals, so that generally their phase relationship remains constant and all is well with repro-duced hues.

If we have any non-linearity in the signal's path, however, a problem can arise. We have seen that the chroma signal proper rides on the luminance signal which can be anywhere between black and peak white levels. The colour-burst, how-ever, is always sitting on the black-level and thus passes through the system at a low level. Any source of timing error which is *level-dependent* will deal differently with the burst and chroma signals, leading to wrong-axis detection and conse-quential hue errors. This effect, known as *differential phase distortion,* can be quite severe, and to counter it, NTSC receivers are fitted with a hue control with which the phase of the subcarrier regenerator can be adjusted to somewhere near the correct axis, as subjectively judged on flesh-tones. This is somewhat haphazard, and certainly inconvenient, as hue adjustment is often necessary on channel-changing.

To overcome the differential phase problem, a modification of the NTSC system was suggested by Dr Walter Bruch and developed by his team in the Hanover, Germany, laboratories of the Telefunken company. This, the PAL system, has been adopted by the UK and some European countries for terrestrial broadcasting with great success. PAL counteracts any phase error which may be present on one line by introducing an equal and opposite error on the next line. The errors are cancelled electrically by an 'averaging' process in a *delay-line* matrix, to be described later, before being demodulated and used to write colours into the display with great accuracy.

Phase Alternation, Line

The scheme is achieved by the reversal in phase (effective inversion) of the V chroma signal for the duration of alternate scanning lines. What happens is that on one line of a field the phase of the V chroma subcarrier is normal, then on the next line of the same field the phase reverses. These can be regarded as 'normal' and 'inverted' lines. While this is happening the phase of the U signal remains normal. *This is not phase-inverted on alternate lines.*

The diagrams in Figure 4.2 show how this corrects the effect of phase distortion. Diagram (a) shows a 'normal' line and phase distortion on a green element causing the phasor angle to lag 45° from the correct 241°. The full-line phasor represents the correct phase and the broken-line phasor (in all drawings) the error phase as received. The green element on this line, therefore, is displayed towards yellow.

Now, on the next line of the field we have the 'inverted' line which is shown in diagram (b). Notice here that the effect of the phase inversion is to invert the diagram and to reverse the direction of the phasor, so that the error now leads the correct 241° by 45°. The green element on this 'inverted' line, therefore, is displayed towards cyan.

Diagram (c) clearly shows how the average phase of the two errors (196° on the 'normal' line and 286° on the 'inverted' line) works out to the correct phasor for green, which as we have seen is 241°. In Chapter 6 we shall see how this averaging process is achieved by means of a glass delay line and an adder network.

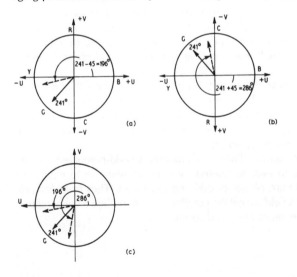

Figure 4.2 *Diagrams revealing in simplified form the phase-error combating artifice of the PAL system. (a) Phasor due to green element occurring at 196° instead of 241° on a 'normal' line. (b) The same phase error on a 'reversed' line. The error is now reversed, and diagram (c) shows that the average of the two errors is 241°, corresponding to the correct phasor angle for green*

V chroma detector switching and synchronising

Perhaps the reader has now become aware that, owing to the V chroma subcarrier phase reversals, the R – Y output from the V chroma detector would also reverse in phase unless some form of compensatory switching were introduced at that particular detector. PAL receivers therefore incorporate an electronic switch to cancel out the PAL characteristic once its usefulness has been realised; it takes the form of an inverter operating on either the V chroma signal or the subcarrier reference to the V detector – either method will achieve the desired result. The switch can be operated by pulses from the set's line timebase which will 'set' and 'reset' the switch on a line-by-line basis. More difficult are the circuit and control required to ensure that the detector switching is synchronised to the 'normal' and 'inverted' lines as transmitted, for clearly if the detector was switched to work from a 'normal' line when the input was an 'inverted' line the displayed colours would be totally wrong! Thus we need to send an identification (*ident* for short) signal to synchronise the receiver's PAL switch.

Swinging burst

This synchronising signal is conveyed by the colour bursts of the PAL signal. It will be recalled that the fundamental purpose of the bursts of the NTSC system is to frequency- and phase-lock the subcarrier regenerator at the receiver (other functions of the bursts include colour-killer switching and automatic chroma control, both dealt with in Chapter 6). The PAL bursts, however, are made to swing in phase 45° either side of the – U chroma axis in synchronism with the phase alternations of the V chroma subcarrier. This is revealed in Figure 4.3, where the burst phase on a 'normal' line is indicated at (a) and that on an 'inverted' line at (b).

Clearly, the *average* phase of the bursts is coincident with the – U chroma axis (180°) and it is this onto which the reference oscillator at the receiver locks, which is the requirement for correct colour reproduction on a PAL receiver.

The NTSC bursts do not swing in phase like this; their phase is fixed to that of the B – Y axis and is constant on all lines.

Figure 4.3 shows the 'normal' lines corresponding to odd-numbered TV lines and the 'inverted' lines to even-numbered TV lines, allowing us to refer the V chroma subcarrier and burst phases to odd and even lines. Thus from *first principles* on, say, line 7 of a field we get the conditions at (a), on line 8 the conditions at (b), on line 9 the conditions at (a) and so on.

Hanover bars

In cases of maladjustment in the receiver's decoder the PAL makeup of the chroma signal becomes visible. If alternately-transmitted colour lines are displayed in different hues due to a phase error, the fact that the TV lines are interlaced, as described in Chapter 1, means that two adjacent lines of the picture

Figure 4.3 *Showing how the bursts swing ±45° relative to the – U axis on successive lines. That is, from 135° on odd lines to 225° on even lines. This means that the average phase of the bursts is 180°, coincident with the phase of the – U chroma axis. Notice how the burst swings are geared to the V chroma phase alternations*

Figure 4.4 *Owing to interlaced scanning, pairs of adjacent lines carry the real and opposing hue errors, and when the error is large this results in the display of Hanover bars*

display carry a hue error, then the next two carry a complementary hue error, and so on, as shown in Figure 4.4. This gives a coarse structure to the Hanover bar pattern, and makes it more noticeable to the eye.

PAL phase identification

Because one PAL switching cycle occupies two lines, the switching frequency is at half line rate, i.e. about 7.8 kHz. This ident frequency is generated in correct phase at the receiver by a circuit which responds to the swinging bursts, and it is this which identifies 'normal' and 'inverted' V chroma signals at the receiver and maintains the PAL switch in correct synchronism. The 'PAL' switch proper is usually a pair of diodes or transistors, internal to an IC, which are alternately switched on by the action of a *bistable*, toggled by line-rate pulses and 'steered' by the ident signal.

PAL encoding system

The block diagram of Figure 4.5 illustrates the basic PAL encoding system. From the camera the RGB signals pass through a gamma correction system into a matrix, wherein the Y signal is derived according to the formula given earlier. This signal passes out of the top of the matrix block in our diagram, then through a delay line on its way to the adder.

This ensures that the luminance signal emerges from the adder in time-coincidence with the chroma signals. Without this delay the luminance signal would arrive early, because the signal is delayed less as the bandwidth of a circuit is increased, and the Y signal occupies a greater bandwidth than the low-definition chroma signals. A similar delay is present in the luminance chain at the receiver for the same reasons, and must not be confused with the chroma delay line!

Returning now to the colour-difference signals, these emerge from the matrix and are weighted before passing into the chroma modulators, each of which is supplied with subcarrier signal; the U modulator direct from the subcarrier generator, and the V modulator via a 90° phase shift (to set up the quadrature condition) and the PAL switch (to impart the alternate line inversions). Modulator outputs are combined in an adder where the burst signal is also gated in, coming from the subcarrier generator via a gate (opened during the appropriate section of the back porch) and a switched phase-shift network to swing the burst phase through a total of 90° line by line, as per Figure 4.3.

We can see now that in essence the PAL variant differs from the NTSC signal in that alternate line phase inversions of the R – Y subcarrier, and synchronised

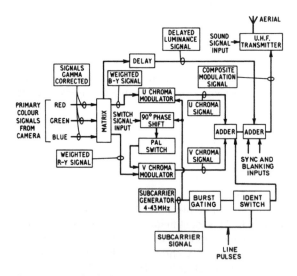

Figure 4.5 *Block diagram of basic PAL transmission system. Note that U and V modulators are balanced types, which means that output from them is only present when both inputs are present*

swings of the burst phase, are incorporated as a method of facilitating the cancellation of signal phase errors and hence hue error at the receiver. Let's sum up the PAL system as we have discussed it so far. The luminance signal is transmitted within the parameters of a conventional monochrome transmission. The chrominance signal is carried in the sidebands only of a subcarrier which is suppressed at the encoder; this suppressed subcarrier carries hue information by virtue of its phase and saturation information by virtue of its amplitude. To facilitate demodulation the subcarrier needs to be regenerated at the receiver, and a reference burst is transmitted once per line on the luminance back porch to synchronise the subcarrier regenerator. The V signal is phase-inverted on alternate lines to give a phase-correction-by-averaging feature, and these 'inverted' lines are identified by the swinging characteristic of the burst phase. The whole shooting-match is then sent to the transmission point, where it is reunited with the studio sound signal for transmission to the viewers' homes.

From PAL to MAC

The original NTSC colour-coding system was devised around 1950 and its PAL variant adopted about 15 years later, as was another form of colour encoding, SECAM. The latter was developed in France and is now in use in that country, much of Eastern Europe, and others. All these systems have the primary objective of compatibility, and all achieve it by some form of spectrum-interleaving of the Y and chrominance signals. We have seen how, in the 625/PAL mode, the colour subcarrier frequency is carefully chosen to ensure that luminance and chrominance sidebands interleave.

What problems are introduced by this spectrum-sharing of the Y and C components of the colour TV signal? So far as the viewer is concerned, the main one is the introduction of *cross-colour*, in which luminance signals at frequencies near the subcarrier are accepted by the decoder and interpreted as chroma signals. They are demodulated to produce random colour patterns superimposed on the fine-detail areas of the picture, and in subjective viewing are most noticeable on televised objects like an actor's striped shirt or a distant picket-fence, and particularly the finer definition gratings of a test card. Cross-colour manifests itself as random herringbone patterning in red/green and blue/yellow. The effect is graphically demonstrated in the two photographs of grating-patterns in Figure 4.6. On the left is an off-screen photograph of a green test pattern embracing virtually the entire luminance bandwidth. The photograph on the right shows the display after passing this pattern through a conventional PAL *codec* (coder/decoder) system. Interference patterns are very obtrusive and the 'beat' effects, centred on 4.43 MHz, can be clearly seen.

There are several other defects inherent in the spectrum-sharing scheme such as cross-luminance (a luminance dot-pattern adjacent to chrominance changes) and the relatively poor chrominance definition we have already discussed – most noticeable on small captions, where horizontal bars of letters are coloured while vertical bars are colourless. The latter defects are not too objectionable on small screens, but would become so if and when a move is made to larger displays. Some solutions to these problems are discussed in Chapters 10 and 11.

Figure 4.6 *Cross-colour: the interference patterns resulting from band-sharing by luminance and chrominance signals. On the left, the signal in green as generated; on the right the display from a PAL codec system*

5 *Transmission and reception*

At present most of our programmes come to us via the very large network of terrestrial UHF colour transmitters which has been built up since 1967 when the first UHF/colour broadcasts started on BBC2. In 1969, UHF colour broadcasting on BBC1 and ITV began, to be joined by Channel Four in 1982. In the early days, there were only a few *main* transmitters, each centrally sited in its service area, and radiating at high power. Secondary and relay stations followed, providing service in 'shadow' and poor reception areas. Today there are about 1000 UHF TV transmitting sites in the UK, each radiating four channels at ERP (effective radiated power) levels ranging from 1000 kW at Crystal Palace, London, to 0.7W at Glyncorrwg in Wales. The effective coverage of these ground stations is well over 99 per cent of the population, a remarkable achievement when it is considered that UHF waves get little further than line-of-sight!

UHF bands and channels

The terrestrial TV-transmitting spectrum is divided into two sections: Band IV containing channels 21 to 34 from 471.25 MHz to 581.25 MHz, and Band V containing channels 39 to 68, 615.25 MHz to 853.25 MHz. The sound carrier in each channel is spaced 6 MHz above the vision carrier. Each local group of channels works within a given UHF spectrum, usually 88 MHz wide. Some groups have a wider spectrum, and each group is known by a letter and colour, mainly to facilitate aerial selection, since aerials are designed to have a bandwidth appropriate to the group for which they shall be used. Thus colour TV viewers need only one UHF aerial to receive all four programmes from the *co-sited* transmitters.

Transmitter links

The sound and vision signals from the studio centre are sent over a cable or radio link to main transmitters, and to many secondary transmitters; most transmitters are in high and relatively inaccessible places. The cable link may be a co-axial or glass-fibre wideband system, often provided and maintained by British Telecom for the transmitting authorities. Where a radio link is utilised it consists of a microwave system, with dish aerials at each end, very accurately aligned with each other. The receiving dish can often be seen about halfway up the mast at the transmitting site. Studio signals are directed through a control centre where *network* distribution is carried out; this acts as a 'switchboard' for incoming and outgoing transmissions.

Transmitter range

At UHF the distance over which the signals can be transmitted is limited to a little beyond line-of-sight, due to the curvature of the earth's surface. Thus reliable reception can only be assured within 30-40 km of a high-power transmitting aerial even when no obstructions are present. On a high receiving site 60km or more may be achieved, though weather and propagation variations would probably affect reception.

To 'illuminate' those areas out of range of the main transmitters or shaded by local hills or mountains, *relay* transmitters are used, typically with an ERP of 1kW or less per channel. Most work on the 'transposer' principle, in which the transmissions from the nearest main station are picked up by a suitable aerial, and each carrier is beat against a stable local oscillator to produce a third frequency which is modulated in the same way as the incoming transmission, and will form (after filtering and amplification) the 'relay output signal' for local use. This avoids the necessity to demodulate and subsequently remodulate the baseband signals, and makes for good economy and a high standard of rebroadcast signal.

Polarisation and ERP

When UHF signals are launched from a transmitting aerial, the geometry of the latter can be designed to impart either a vertical or horizontal polarisation to the electric field created in the air. By suitably polarising transmitting and receiving aerials, a great reduction in the level of interference between transmitters working on the same channel can be achieved, and this property is fully exploited in arranging the 4000 or so (1000 x4) transmitters in the UK to share the 44 UHF channels available with a minimum of interference, even in the face of tropospheric abnormalities which tend to increase the transmission range. A precision *offset* of carrier frequency is also employed to minimise the subjectivity of any co- or adjacent-channel interference which may occur.

We have referred several times to ERP in discussing transmitters, and the effective radiated power is that which is actually *radiated* by the aerial. Because transmitting aerials are designed to have a *directional* field, the suppression of radiation at some angles allows a greater concentration of energy in the wanted direction, so that a more concentrated beam is achieved, with a useful gain in transmitter range. Because all reception takes place at ground level, upward radiation into space is suppressed; many transmitters have horizontal directional properties, too, so that wasteful transmission over the sea or into a mountainside is avoided. By this means the ERP takes account of the directional power gain of the aerial system. This 'gain' characteristic is a feature of receiving aerials, too, as we shall see later.

Vision transmitter

Each station usually has a standby transmitter which comes into operation should the main one fail. In main transmitters the sound and vision signals are dealt with separately until they are in high-power RF form, when they are routed into a combining unit, from which a single co-axial cable takes the UHF signals to the common radiating aerial.

Some large transmitting stations include basic studio equipment such as slide scanners and caption generators for producing test signals or 'breakdown apology' captions. The broadcasters maintain a number of ROCs (Regional Operations Centres) at main transmitter sites, from which all the main and secondary transmitters in a region are monitored and controlled.

The video signals from the studio centres are received, monitored and then 'clamped'. This latter procedure is very important in television. The various components of the composite signal have to be maintained in specific proportions. The black level intervals, for example, have to hold steady at all times, as does the amplitude of the sync pulses and colour bursts. Variations in proportions and levels can occur in the signal after leaving the studio centre, and so it is arranged that the video signal is clamped electrically to a stable and fixed level before it is applied to modulate the carrier wave. Teletext signals in the field blanking interval are extracted, 'cleaned' and reinserted at this stage, then the baseband signal is ready to pass into the modulator.

As we have seen in Chapter 1, amplitude modulation is used for the vision signal as shown in Figure 1.11. The high-power AM UHF signal is launched into space by a stack of dipoles or slot elements at the top of the mast, each radiating element being spaced by half the wavelength ($\lambda/2$). The aerial's directional properties are imparted by carefully adjusting the phase of the carrier signal fed to each tier of elements. To achieve a downward tilting of the beam, for instance, a progressive phase lag is introduced into the carrier as we move downwards from the topmost tier, and in practice this is done by suitable trimming of the feeder cables to each tier. The entire radiating panel is then encased in a closed fibreglass cylinder for protection from the weather – this gives the characteristic 'candle' appearance that we see crowning the transmitting mast.

Sound transmitter

The sound signal is also brought into the transmitting station and monitored before being fed to the preamplifiers which supply the modulation input to the sound transmitter. In the UK TV system, terrestrial transmitters use FM for sound, which has several advantages over AM. The dynamic range of the audio information (e.g. the range between the softest and the loudest sounds) can be more readily retained, and because most of the interference experienced on sound reception (and indeed vision) results from amplitude variations of the modulated carrier wave, this can be eliminated by amplitude *limiting*, that is shaving off the top and bottom of the waveform at the receiver, when the sound is carried by FM. Moreover, high-quality sound reproduction is facilitated by an FM system, and compatible stereo is possible by several different techniques. The UK has settled upon the *Nicam 728* system for stereo TV sound. Here an additional carrier at + 6.552 Mhz carries pulse-code modulation signals containing information on 'L' and 'R' sound channels, quite separate from the established FM carrier (see Chapter 12).

Other desirable features stem from the use of FM sound in the system as a whole. For example, it makes possible at the receiver a common IF (intermediate-frequency) channel for the sound and vision signals, owing to the lack of interaction between the two types of modulation. It also facilitates the system of *intercarrier sound*, where a 6 MHz FM sound signal is produced from the beat between the sound and vision carriers, spaced as they are by exactly 6 MHz. Since the accuracy of this 6 MHz signal is ensured by the crystal-controlled references at the transmitter, it follows that the intercarrier sound scheme eliminates the effect of any tuning-frequency drift at the receiver on the sound signal – it would be far otherwise if the sound IF at 33.5 MHz were directly demodulated.

Because of its lesser bandwidth and the FM modulation system there is no need for the sound transmitter to have so great a power as its vision counterpart. In system engineering, the ideal is that in circumstances of signal attentuation, interference etc., all components of the system should have about the same threshold of failure; there is little point in maintaining a perfect sound signal when the picture is completely obliterated by 'snow' or synchronisation is lost in the presence of heavy interference. This was a guiding principle in the design of the colour-encoding system and sync/vision ratio. For sound, it has been found that the 'common threshold' is achieved when the ratio of vision to sound carrier power is about 10:1 (sound carrier 10dB below vision carrier power), and this is commonly adopted for UHF TV transmitters throughout the UK.

Vestigial sidebands

We have seen in Chapters 1 and 3 that the information transmitted by a modulated carrier wave lies in both the sidebands (see Figure 5.1a). Each sideband carries the same information (they are mirror-images of each other) and so a duplication is present, and more bandwidth than necessary is used up. It is possible to send a complete signal using one sideband only; and this is necessary in the UHF bands to conserve precious spectrum space. However, it is not possible (by filtering) to sever one sideband as cleanly as shown at (b) in Figure 5.1, so suppression usually takes the form of a 'tail' or a gradual roll-off of the unwanted portion. The choice is to have this occurring either in the wanted sideband or in the unwanted one, as shown by (c) and (d) in Figure 5.1. Neither of these alternatives is really satisfactory. In the first case low-frequency energy is lost if the roll-off occurs in the wanted sideband, while in the second case incomplete suppression takes place and the bandwidth of the signal as a whole is still fairly large. In

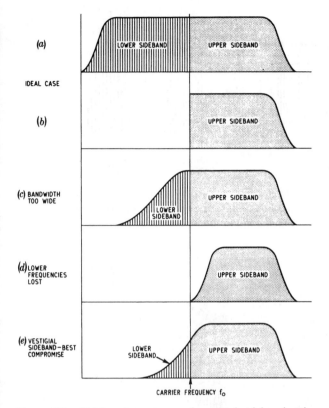

In the figure:

(a) IDEAL CASE — LOWER SIDEBAND | UPPER SIDEBAND

(b) — UPPER SIDEBAND

(c) BANDWIDTH TOO WIDE — LOWER SIDEBAND | UPPER SIDEBAND

(d) LOWER FREQUENCIES LOST — UPPER SIDEBAND

(e) VESTIGIAL SIDEBAND – BEST COMPROMISE — LOWER SIDEBAND | UPPER SIDEBAND

CARRIER FREQUENCY f_o

Figure 5.1 *Sideband structure and vestigial sidebands. These are explained in the text*

television a compromise is sought, where the lower sideband is partially suppressed, and because there is a vestige of it remaining, the scheme is known as *vestigial sideband transmission.*

The net effect of this type of transmission is that the TV transmission uses the practical minimum of band-space, but when picked up and processed by a receiver suitably designed to accept it the asymmetrical sideband energy is compensated for and equalised to give reception as good as that from a double-sideband system. The spectrum of energy for a single transmission channel is given in Figure 5.2, where the sound and vision carriers can be seen with sideband energy distribution and limits. This diagram also shows the interleaving effect of the luminance and chroma components of the vision signal within the upper sideband of the vision carrier.

It should be borne in mind that this is but a single 8 MHz-wide channel in the broadcast spectrum. One has to imagine the diagram of Figure 5.2 reproduced many times, with each laid side by side, stretching to right and left of our diagram, to get a picture of the UHF band in action. Obviously, receiver bandwidth must be not only carefully shaped in its response curve, but also provision must be made to reject (by means of notch filters) out-of-band signals in adjacent channels.

Figure 5.2 *Detailed diagram showing how the sidebands of the chroma signal are interleaved with those of the luminance signal. The diagram also shows the sidebands of the sound signal and the suppressed lower sideband*

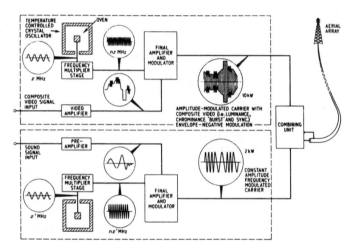

Figure 5.3 *The prime elements of a terrestrial colour television transmitting station*

To round off our description of terrestrial transmitters, the diagram of Figure 5.3 shows in simplified form the basics of a broadcasting set-up.

UHF reception

Having launched our carefully composed RF carrier signals into the air, it's now time to examine the means of recapturing them at the receiving site and turning them back into an electrical signal for application to the TV set. Like any radio receiver the TV tuner has the job of selecting one channel from the many presented to it, and rejecting others which are out of channel. The conventional UHF aerial seen on millions of rooftops is the *Yagi* type, named after its Japanese inventor, through other types are sometimes used.

Aerial Bandwidth

Aerial elements are usually referred to in terms of wavelength, and a dipole is said to be 'tuned' when its length corresponds to one-half of the wavelength of the carrier it is to receive. Because the wave velocity reduces slightly when it is captured by the aerial element, the aerial (physical) half-wavelength is about 95 per cent of the signal half-wavelength.

At the high frequencies of the UHF channels, the active element may not be much more than 200 mm. However, to avoid attenuating the sidebands of the wanted colour signal, the aerial must have a sufficient bandwidth to pass not only the signals of one channel, but also those of the other three in the local group. As we have seen, the local group of channels may extend over 88 MHz or more. If the aerial bandwidth is insufficient or if the aerial is of an incorrect channel grouping for the area, chroma sidebands may be seriously attenuated to give severe desaturation; more often one or more channels of the group may have a poor S/N ratio due to uneven response of the aerial. The standard colour-codes for receiving aerial groups are given in Table 5.1.

Table 5.1. Groupings and colour codes for UHF receiving aerials

Group	A	B	C/D	E	W
Channels	21–34	39–53	48–68	39–68	21–68
Colour code	Red	Yellow	Green	Brown	Black

The Yagi array

The most popular type of TV aerial to date is the Yagi design, developed in the 1920s and much used since then for its good gain and directional properties. It consists basically of a half-wave dipole, one or more reflectors and a large number of directors, as shown in Figure 5.4. When the array is used for reception the signal energy is 'focused' on the dipole (which of course is the active element) by the directors, and any which disperses beyond it is collected and re-radiated back to the dipole by the reflector behind it. The net result is an array with high gain (compared to a single dipole) due to its highly directional property which also confers an ability to reject signals which arrive off-beam, useful in areas like the UK where many transmitters are present, working on the same or adjacent channels, and where buildings and natural topography can cause signal reflections.

Whether an element is to act as a reflector or director depends on its length relative to the dipole. When it is longer it behaves as a reflector and when shorter as a director. The element spacing on the boom in terms of wavelength is also of

Figure 5.4 *UHF Yagi array set for receiving a signal of horizontal polarisation*

importance in determining the behaviour of an array. The gain tends to increase as more directors are added, but the law of diminishing returns eventually comes into play, and 18 elements are the normal maximum in domestic aerials. If more gain is required, two or more aerials can be used together, linked to each other, and the receiver, by a *phasing harness*. The characteristics of this, and the relative spacing of the aerial arrays, can be engineered to achieve specific requirements of gain and directivity.

The reflector and director elements are termed 'parasitic', meaning that they are not receiving dipoles as such. As more elements are added, so the centre impedance of the dipole falls, and in order to maintain a reasonable match to the feeder the dipole is folded double as shown in Figure 5.5. This artifice steps up

Figure 5.5 *A folded dipole connected to a co-axial cable feeder. Notice the reflector behind and the director in front, relative to the direction of the transmitter*

its centre impedance by four times and thus compensates for the presence of multiple parasitic elements.

Aerial installation

While set-top and loft aerials may give passable reception in clear zones within the primary service area of a transmitter, the only reliable installation is a roof or chimney-mounted aerial with a low-loss feeder to the TV receiver. The elements must agree with the polarisation of the transmitter to be used, which generally (but not always!) means that they must be horizontal for main transmitters, vertical for relays. In situations of low signal strength, a slight upward tilt of the front 'firing' end of the array may improve results by a dB or two, though a *masthead amplifier* may be necessary to achieve an adequate S/N ratio in the picture in very poor reception areas. This consists of a small wideband transistor amplifier mounted very near the dipole to amplify the signal before further degradation occurs in the feeder cable, which now carries a DC power supply for the masthead amplifier, as well as the UHF signals on their way to the receiver.

Normally the aerial needs to be 'looking' directly at the transmitting mast, and this is generally achieved by swinging it while monitoring signal strength on a suitable hand-held meter. Sometimes, due to multipath reception and the resulting 'ghosting' effect on pictures, an orientation must be chosen for minimum spurious pickup rather than maximum signal strength.

Log-periodic aerials

An alternative to the Yagi array is the log-periodic type, which works on a different principle, although it looks superficially similar. It consists of a pair of parallel booms mounted a few centimetres apart with separate dipoles mounted at intervals along its length, their connections being transposed between adjacent elements. At various frequencies in the UHF band different dipoles come into tune due to their differing lengths, while the twin beams on which they are supported act as a transmission line, at the end of which the signal is matched into a co-axial cable. The elements adjacent to the active dipole (whichever it may be at the frequency in use) act as directors and reflector. Thus the log-periodic aerial has some directivity, but not the very narrow beamwidth of a Yagi array of comparable size; its gain is also comparatively low. The main advantage of the log-periodic type is its superior bandwidth and absence of spurious resonance effects. As such, it is useful as a measuring aerial (the broadcasters use them to plot field-strength contour maps), and in areas where a wide spread of grouped channels is encountered. They are also popular for use on caravans and boats where (hopefully!) reflections and signal strength problems will not be encountered, and one aerial can be expected to operate over the whole of UHF bands IV and V.

Down-lead and matching

The characteristic impedance of the dipole of an aerial system is designed to appear as close as possible to a 75R resistor. This matches the impedance of the transmission co-axial cable which conveys the signal to the receiver's aerial socket, ensuring least loss from standing waves due to mismatching. At UHF the attenuation in the down-lead is significant and low-loss feeder is recommended to minimise this, installed to avoid sharp bends or kinks. The co-axial cable itself consists of a flexible outer sheath and an inner conductor supported through the centre of the sheath by low-loss insulating material, called the dielectric. This type of feeder has a characteristic impedance of about 75R. The half-wave dipole in a Yagi array has its ends terminated in a weatherproof box, wherein the feeder is connected, sometimes via a small *balun* or matching transformer. In the absence of this the inner core of the co-ax cable connects to the upper dipole end while the outer sheath is bonded to the lower dipole end.

For further information on aerials, siting and transmitters in the UK, consult the Engineering Information Department of the BBC (201 Wood Lane, London, W12 7TS, tel. 0181 752 5040), or the ITC Engineering Information Section (tel. 01962 848647).

6 *Colour decoding*

Since we discussed the PAL encoding system we have strayed into RF propagation and amongst the chimney pots. Readers who feel it necessary are advised to run over Chapters 3 and 4 again before continuing here with the decoding or restoring process.

Let's first define what we are seeking to achieve. We left the chroma signal, in PAL form, riding on the luminance waveform as a modulation signal for the transmission system. This modulation signal will be rendered in baseband form at the receiver's vision detector, and the job of the decoder is to process this signal to derive R, G and B drive signals to drive the display device. Here we must digress for a moment into the terminology we are using. In the original colour TV receiver plan (at the outset of colour broadcasts) the *decoder* dealt only with chroma signals and its product was the three colour-difference signals, R – Y, G – Y and B – Y. The final matrix for RGB was the picture tube itself, which summed the two signals applied to each of its guns – difference signals to the grids and the Y (luminance) signal to the cathodes, the latter sourced from an independent video amplifier just like that used in the monochrome set. Thus in some set designs the decoder could be removed from the set altogether, leaving it operating in monochrome on the Y signal. For many years now the colour-processing circuits of a TV have dealt with both Y and chroma signals together to render R, G and B signals as output; three separate primary-colour video feeds to the display device, in just the same form as they emerged from the three separate image sensors of our Figure 3.3. This is the current concept of a decoder (in the context of a PAL signal, anyway), and throughout this chapter this convention will be used.

Taking first the reference (burst) signal, we need to use this to establish within the decoder a reference timing 'clock' against which to measure the timing variations of the chroma signal itself – it will be recalled that the timing of the subcarrier is indicative of the hue being described. We need to provide an inverting switch to restore normality to the V signal on 'PAL' lines; and control it by means of the swinging alternations of the burst signal. In the chroma signal chain is provided the delay line matrix which overcomes the effects of differential phase distortion. During demodulation the subcarrier is discarded, then the baseband V and U signals are de-weighted and matrixed to derive a G – Y signal. Finally, all three colour-difference signals are added to the Y component to give low-level primary-colour signals. Further amplification brings these up to the level required by the picture tube or other display device.

Chroma channel

The chroma channel is contained in the block diagram of Figure 6.1. It consists of the two chroma amplifier stages, with the ACC (automatic colour control) and colour-killer connections, the delay-line driver, the PAL delay line and the PAL matrix. The other sections, though associated with the chroma channel, will be considered later.

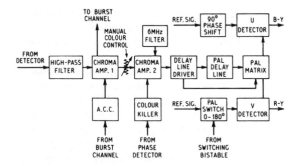

Figure 6.1 *Block diagram of chroma section of colour decoder. Each section is fully explained in the text*

Now the composite signal from the vision detector, assuming that the receiver is correctly responding to a colour signal, is passed to chroma amplifier stage 1 through a high-pass filter. Since the chroma information is centred on 4.43 MHz, the high-pass filter accepts all this information while attenuating the lower-order components constituting the luminance information. Thus in essence only chroma information arrives at the input of the first chroma stage. The signal is then amplified by the second chroma stage and by the delay-line driver, these having a bandpass characteristic of about ±1.2 MHz centred on 4.43 MHz. The skirts of the response show a fairly steep roll-off at either side, determined by the filtering in the chroma signal path to the input. Manual control of the level of the chroma signal is provided by the user colour control, while automatic control (ACC) is also provided at the first chroma stage. The second stage in the block diagram is controlled by the colour killer.

The ACC is generally applied to the stage as a potential which regulates its biasing and hence gain depending on the level of the chrominance signal, as indicated by the amplitude of the colour-bursts (the only part of the chroma signal transmitted at a constant, known level). This reference level is picked up from the burst channel (described later) and then rectified to yield the control potential.

The second chroma stage in Figure 6.1 is deliberately designed to be non-conducting (e.g. biased off) in the presence of monochrome signals. This allows the receiver to work in black-and-white without the colouring circuits being active. If they were open during this time colour interference and noise might mar the monochrome display due to spurious colouring signals getting through the chroma channel. Now, when the colour killer detects a colour transmission by the presence of bursts, it produces a 'turn-on' bias which opens up the chroma channel, thereby letting the legitimate colouring signals through. The colour killer generally receives ripple signal (at half-line frequency; see later) from the burst-phase detector and this, or a part of it, is rectified and smoothed to produce the switch-on bias for the controlled chroma stage. A final function of the second chroma stage is that of *burst blanking*. We do not need the burst signal in the chroma chain beyond this point (it has already been diverted to the reference signal channel) and a line-rate pulse, suitably delayed, is fed to the second chroma stage in time-coincidence with the burst signal to blank out the latter and prevent it from upsetting functions further downstream in the signal path, such as *clamping*.

The delay-line driver thus delivers chroma signal to both the PAL delay line and the PAL matrix, and it is this part of the chroma channel which deletes the 'phase sensitivity' of the chroma signal, and neatly separates the V and U signals for feeding to their respective demodulators.

The PAL function

We have already dealt with the PAL function from the signal point of view in Chapter 4, but since it is particularly important no apology is made for referring again to it here, but this time more specifically from the point of view of the receiver. The relevant circuit section of the PAL delay line and matrix is given in Figure 6.2. Some sets employ an arrangement just like this, while others differ in detail; nevertheless the net result is the same in all cases.

Chroma signal is driven into the delay-line input transducer which converts the electrical signal into an acoustic one, whose rate of propagation through the glass block is relatively slow. Mechanical vibrations travel along a defined path within the glass block, whose edge surfaces are arranged to fold the path by multiple

Figure 6.2 *The delay line and its matrix. The folded accoustic path within the delay line is arranged to take 63.943 μs to traverse*

reflections; see Figure 6.2. The delay suffered by the acoustic wave in this long folded path is just under the period of one TV line, after which it arrives at the output transducer and is reconverted into an electrical signal by what amounts to a piezo microphone on the receiving face of the glass block. Chroma signal is also picked up from preset P1 and from this fed to the output of the line. The output transformer T2 thus receives two lots of chroma signal, one delayed in the line, and one direct. It is the job of T2 to add and subtract (i.e matrix) direct and delayed lines of chroma signal. These processes, it will be recalled, effectively remove the 'phase modulation' from the chroma signal.

It works like this. We get addition and subtraction because the transformer T2 is bifilar wound and centre-tapped. The effect is that the signal from the delay line across winding A is exactly 180° out of phase with the signal also across that winding from preset P1, while the signal from the delay line induced into winding B is exactly in phase with the signal also across that winding from preset P1. The antiphase signals across winding A thus cancel out, assuming equal strengths (one is subtracted from the other), while the in-phase signals across winding B add together.

Let us first consider the action on the U components of the direct and delayed signals. The subcarrier upon which the U signal is based is not phase-reversed at any time in transmission, so it has the same base phase on every line. Since the delay is engineered to be 63.943 µs it holds $283\frac{1}{2}$ cycles of reference subcarrier, and the emerging signal is half a cycle out of phase with the direct one – this represents a phase relationship of 180° so that when direct and delayed U signals are *added* they will cancel out thus: $-U + U = 0$. From the subtracting function, we get: $-U - U = -2U$, assuming equal strength direct and delayed signals of unity value. The left-hand side of winding A on T2 in which subtraction occurs is thus endowed with a 'pure' U signal (albeit negative) and no V signal at all.

The V components of the direct and delayed signals are similarly processed, but the net result is different because (see Chapter 4) the transmitted subcarrier upon which the V chroma is based is reversed in phase during alternate lines, which means that the V components themselves are similarly processed. The adding part of the matrix thus 'sees' V signal of, say, 'positive phase' via the delay line and V signal of 'positive phase' via the direct route. If this is not clear, remember that $283\frac{1}{2}$ cycles in the delay line gives a phase inversion between input and output. With the transmitted signal having also undergone a phase inversion since the last line (PAL feature), the two inversions cancel out *on every line* so that the adder (T2 section B) will produce $V + V = 2V$ on every line, while the subtractor, section A, will always produce $V - V = 0$. So it is that the right-hand side of T1 section B will produce pure V signal with no cross-talk from the U channel.

To summarise, therefore, the U detector receives only U chroma signal because the alternate-line phase reversals of the V chroma signal result in cancellation of the V chroma components in the subtractor section, while the V detector receives only V chroma signal because the in-phase V components reinforce in the adding section while the antiphase U components are added and thus cancelled. In other words, the composite chroma signal is very neatly separated into its two V and U constituents, these now carrying information in terms of amplitude *and not phase*. This means that phase distortion (which is tantamount to 'timing errors') in the system can no longer introduce hue errors on the display. If differential phase errors are fairly large there is a mild by-product in the form of reduced saturation, but this is more subjectively tolerable than a change of hue!

We still need to introduce the reference signals to the V and U chroma detectors (their operation was described in Chapter 3 and illustrated in Figure 3.17), in phase quadrature of course, to recover the quadrature-modulated baseband V and U signals. We must also neutralise the effect of the ±V chroma signal at some stage. In Figure 6.2 P1 adjusts the matrix balance so that the direct and delayed signals have equal amplitude. Imbalance here can give rise to Hanover bar effects as described in Chapter 4.

We can now return to Figure 6.1 again and see that the reference signal to the U detector undergoes a 90° phase shift which provides the quadrature condition just noted. It may be recalled that in our simple illustration of the delay line matrix system we showed the U output as being in *negative* form. (– 2U at subtractor output) and to correct this we would need to either invert the U signal before detection or *introduce a further 180° phase shift in the reference signal to the U demodulator. Either will achieve the desired effect of rendering, at the U detector output, a correct-polarity signal. In the case of the V-detector we have to neutralise the phase inversions of the transmitted signal, and as described for U above, this can be achieved by either inverting the V signal itself, or the subcarrier feed to its detector, but this time on a line-by-line rather than a permanent basis. The switch thus operates effectively at line timebase frequency, but remember that this is the 'toggling' rate; Figure 6.1 shows the switch in the reference signal path.

The PAL switch commonly consists of a pair of diodes or transistors within an IC, switched alternately to opposite phases of the local reference signal, the actual switching being governed by a bistable generator (this is of the multivibrator family but is stable in both switching modes and requires a trigger pulse to switch it from one state to the other) triggered by pulses from the line timebase or sync separator. It thus switches state line by line, each state having a frequency equal to half the TV-line repetition frequency.

Provided the switching count corresponds to the plus and minus phases of the V chroma signal, therefore, the V detector will work as though it is receiving a V chroma signal of constant base phase; but if the switching count is in error the V detector will fail to work correctly and the displayed hues will be in error. To ensure that the switching is correctly synchronised to the transmitted V chroma phase inversions, line-identification pulses are also fed to the bistable from the phase detector responding to the swinging bursts, but this story is yet to come.

Burst and reference signal channels

Another section of the PAL decoder is given by the block diagram of Figure 6.3. Here the burst gate and amplifier receives composite chroma signal from the chroma channel. The gate part is a transistor or diode which is switched at line frequency such that it conducts only during the periods of the bursts; during the rest of the television line it is non-conducting, thus performing the opposite function to the burst-blanking switch in the chroma amplifier described earlier. This means, then, that only the bursts of the chroma signal are amplified and that the output from this 'block' consists only of a series of bursts with no picture content.

Note that the bursts are also fed to the ACC circuit in Figure 6.1 which, after rectification and smoothing, provide the control bias for the chroma amplifier, already considered. There are many possible causes of chroma-signal level variation, but the ones we are most concerned with are those where the chroma signal is affected to a different degree than the luminance component. The resulting imbalance would upset saturation if it went uncorrected by some form of ACC; the most common causes of this sort of imbalance are RF mistuning and aerial or propagation defects which unduly emphasise one section of the RF channel bandwidth.

Returning to Figure 6.3 and the bursts, however, they are primarily fed to a phase detector which compares their average phase (remembering that their 'swinging' characteristic takes them alternately ±45° relative to the – U axis; see Figure 4.3) with the phase of the locally-generated reference signal. This reference signal for such phase comparison is fed via a 'sample loop.'

Figure 6.3 *Block diagram of reference and burst channels*

The reference generator is crystal-controlled but mild phase variation of the reference signal is made possible by a varicap diode effectively in shunt with the crystal. The varicap diode is widely used in applications where a DC-controlled reactance effect is required. It consists of a p-n semiconductor junction operated in reverse-bias which, under this condition, exhibits a capacitance effect; the effective value decreasing as the reverse bias is increased. This happens because the *depletion layer* between the heavily doped p and n regions acts as a dielectric and its width is a function of applied reverse bias. Thus we can vary the crystal's shunt capacitance, and hence its phasing, by simply applying a DC voltage to the varicap device.

Now, for correct chroma detection it is essential that the phase of the reference signal applied to the detectors corresponds exactly with the phase of the subcarrier which was suppressed at the transmitter. The 90° phase shift required between the two detectors is provided by the phase shifter in Figure 6.1, which has already been mentioned.

The phase detector

It is the job of the phase detector to provide a control potential for the varicap diode such that the phase of the reference signal is always held constant. This is easily possible because the average phase of the bursts corresponds exactly to the average phase of the subcarrier at the transmitter. The phase detector, in fact, produces a DC output of a positive or negative value corresponding to the phase error between the bursts and the locally-generated reference signal. This is called the *error* voltage, and it is superimposed upon the standing reverse bias applied to the varicap diode so that the phase of the reference signal is continuously held at the correct value, with any tendency to drift being immediately corrected. Thus is set up a phase-locked loop (PLL) in which the output signal is locked or *slaved* to the incoming reference, in this case the colour bursts. Modern electronic equipment makes much use of PLLs.

The correctly-phased reference signal is then fed to the V and U chroma detectors via the sections shown in Figure 6.1. Some designs will have a DC amplifier between the phase detector and the varicap diode, while there is always a 'buffer' stage (probably an *emitter-follower*) between the reference generator and the chroma detector feeds.

Ident pulses

We have seen (Chapter 4) that the bursts are swinging in phase line by line such that on one line the phase relative to the – U chroma axis is + 45° and on the next line – 45° and so on. These swings are geared to the phase reversals of the V chroma signal to provide a facility for the identification of the + V and – V lines (odd and even lines, Chapter 4) of the chroma signal, thereby allowing the PAL switch to be correctly synchronised.

So far as the phase detector is concerned the swings of phase are processed as phase modulation, which means that the phase detector's output contains, super-imposed on the error voltage, a ripple at half-line frequency (about 7.8 kHz) – *half* line frequency because the phase is the same on every *other* line.

This ripple can be amplified and squared-off to produce ident (short for identi-fication) pulses which are fed to the PAL switch bistable. It is arranged that when the line-frequency triggered bistable is correctly synchronised (and there is only a 50/50 chance at the outset!) the ident pulses will have no effect, but if the bistable is out of agreement with the transmitted V chroma the ident pulse will inhibit its operation for one count, so that it 'misses a beat', as it were. Thereafter the bistable phase will be correct and the ident signal will not be required until a new channel is selected or the set switched off and on again.

Video section

Having arrived at colour-difference signals once more (at the output terminals of the V and U detectors), we are now ready to process them towards the production

Figure 6.4 *The 'back-end' of the decoder showing the stages of signal processing between the chroma detectors and the picture tube*

of RGB signals. The first step is amplification in the R – Y and B – Y preamplifiers whose gains are adjusted to 'de-weight' the signals as described in Chapter 3, and the relative gains are 1.140 for R – Y and 2.028 for B – Y. Next comes the matrixing section for G – Y (see Figure 6.4) and a *clamping* process. This is similar to that described earlier for TV signals on arrival at the transmitter, and ensures that all three signals, R – Y, G – Y and B – Y are sitting at the same reference level so far as their black level is concerned. This is essential for the next stage, which is the coming-together of the luminance and colour difference signals to form primary-colour signals.

Let us see how this final process is done. Again it is a simple algebraic function in which the Y signal is *added* to each colour-difference signal, having itself undergone a similar clamping process. So (R – Y) + Y = R, (G – Y) + Y = G and (B – Y) + Y = B. When the difference and luminance signals are added, then, we are back to primary-colour signals. For the standard colour bars of our Figure 3.16 the primary-colour waveforms are shown in Figure 6.5. In Chapter 4 we saw (Figure 4.5) that to achieve time-coincidence between narrowband chroma signal and wide-band luminance signal at the transmitter a delay line was inserted in the Y channel. At the receiver's decoder a similar delay is required to prevent the chrominance being printed slightly later (and hence to the right, as TV picture scanning is from left to right) than the corresponding luminance picture.

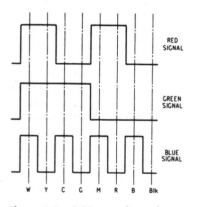

Figure 6.5 *RGB waveforms for a standard colour-bar signal. These result from matrixing Y and individual colour-difference signals*

The delay required is of the order of 300–800 ns depending on circuit design, and the wideband delay line can be seen in Figure 6.4. It is made of series L and parallel C components, rather like a π filter; its cut-off frequency, however, is way above the highest luminance frequency in the video signal.

At the point of emergence from the matrix the RGB signals are at low level, perhaps 4-5V black to 'white' and we shall wish to drive a colour picture tube with them. To provide the 90V or so drive required by the picture-tube guns, each primary colour signal is passed through a wideband amplifier, very often a complementary type using two or more transistors in each channel. The characteristics of the three primary-colour output amplifiers must be closely matched in terms of gain, black level and bandwidth to ensure that the primary colours agree or 'track' at all brightness levels and frequencies. This will achieve correct relative and combined colour fidelity at all brightness, hue and saturation levels.

An IC decoder

The entire PAL decoder function is contained within a single IC (integrated circuit) or 'chip' in current receivers. These use a mixture of analogue and digital functions and require a minimum of external components. Typical of such a decoder chip is the 24-pin plastic-packaged Mullard TDA3562A illustrated in Figure 6.6. In broad outline, it follows the block diagram and description already given, but several detail differences and new features will be noticed, and these will be described in turn.

Luminance chain

The delayed luminance input (pin 8) is AC-coupled to permit clamping within the IC. An amplifier and line-rate black-level clamp is first encountered, then, followed by insertion of a reference black level generated within the chip. The insertion period (3L) occurs once per TV *field* for reasons which will become clear later. Luminance signal then passes directly to the R, G and B matrices, and there we will leave it for the moment.

Chroma path

Chrominance signals enter the chip on pin 4, sourced from the same point (vision detector) as the luminance signal described above; in place of the latter's wideband delay line, they will have come via a high-pass filter, and so will consist of the baseband signal spectrum between about 3.3 and 5.5 MHz to embrace the chroma signal bandwidth. The first step is the ACC function, carried out in the controlled chroma amplifier – controlled, that is, by the potential derived from a peak detector and stored in the capacitor on pin 3. Next comes a gated saturation control stage under the influence of a DC control voltage from the user saturation control. The 'gated' function is puzzling here. In fact the gating prevents the saturation control from having any effect on the burst signal, which thus passes, at a constant level, though the chroma chain, delay line and all. The gated

Figure 6.6 *Block diagram of the TDA2562A chip (Mullard)*

chrominance amplifier increases the chroma/burst amplitude ratio by conferring about 12dB extra gain during the 'picture chrominance' period of 52 µs. An impedance buffer follows, whence the signal leaves the chip on pin 28 en route for the delay line.

Emerging from the delay line, the separated U and V signals now enter the chip on pins 22 and 23 respectively and are passed to the demodulators. The burst is also present, remember, and this is extracted from the delay line outputs for use in the gated burst detector at the bottom of the diagram.

Reference chain

The reference oscillator in this design runs at exactly twice subcarrier rate – $4.433619 \times 2 = 8.867238$ MHz and this is the nominal frequency of the crystal connected to pin 26. Let us explore the reason for this double-frequency technique. We know that the two chrominance detectors require switching pulses with the quadrature relationship, and ordinarily this would require a 90° phase-shift network of some kind in the reference path to the U detector. If we have a reference signal at $2f_{sc}$ we can halve its frequency by use of a bistable counter rather like the ident flip-flop described earlier. If we arrange two such flip-flops, one counting on negative flanks, the other counting on positive flanks, we shall have two 4.43 MHz squarewaves as outputs from them; essentially these outputs will have a 90° (quadrature) phase relationship as required for the synchronous detectors. Figure 6.7 illustrates the technique.

Back now to the bottom of Figure 6.6. The ÷2 90° shift block contains the bistables just described and one output operates the B – Y demodulator. The second output is inverted on a line-by-line basis in the PAL switch on its way to the R – Y demodulator whose baseband output is thus restored to + (R – Y) on both 'PAL' and 'normal' lines. The gated burst detector operates only during the back porch, comparing the R – Y axis reference signal at 4.43 MHz with incoming burst to render an error voltage at pins 24 and 25, between which are connected a time-constant RC circuit to act as 'flywheel', smoothing out the ripple due to the swinging burst. The smoothed error voltage is applied to an internal varicap diode

Figure 6.7 *Derivation of quadrature drive signals for the chroma demodulators. A crystal oscillator generates a squarewave at 8.86 MHz (2 f_{sc}) and a pair of ÷ 2 bistables, respectively triggered from rising and falling edges, produce two f_{sc} drive signals in quadrature*

to 'pull' the crystal oscillator into lock as described earlier. Thus is the PLL set up, and the demodulator feeds are now slaved to incoming burst and held rigidly on the correct phase vector axes for correctly-timed detection and sampling.

Ident circuit

The operation of the ident detector is a little different in this chip from the established practice. Here we have an H/2 detector to compare the subcarrier output of the PAL switch with the R – Y axis component of the burst signal. Let us look a little deeper into this. We have seen that in a properly set-up delay-line matrix the V and U outputs contain only *amplitude* information, with phase variations already taken care of (page 66). Considering then the effect of the burst signal on the V (R – Y) signal, we can see (Figure 4.3) that on a 'normal' line a positive voltage will appear in the V channel during the burst period, then during an inverted or 'PAL' line a negative voltage will be present during the burst period.

When the H/2 detector discovers a discrepancy between R – Y signal phase and R – Y reference phase, the PAL flip-flop gets a 'reset' pulse from the H/2 detector via the ident stage; this corrects the switching sense and ensures agreement between the PAL switch and the transmitter.

ACC and colour killer

Once the PAL switching phase is correct, the output voltage of the H/2 detector is directly related to burst amplitude and so becomes the feedback signal in the ACC loop, finding its way back to the controlled chroma amplifier in the form of a DC control voltage. In the absence of a burst signal killer action is evoked with the chip, taking place at *two* points in the signal path to ensure that spurious chroma suppression is adequate. The saturation control input is taken low and the synchronous detectors are simultaneously cut off.

Control pulses

For economy of lead-out pins, all control pulses enter the chip on its pin 7, and three basic pulse streams are required. A field pulse is present at 20 ms intervals along with a 12 μs line blanking pulse at 64 μs intervals. During the latter, and coinciding with the back porch, a further pulse, about 4 μs wide, is added at a higher level (hence the *sandcastle* shape of the composite pulse). These are separated within the chip by 'level slicers,' then timed and processed for many different functions as the chip diagram makes clear.

NTSC capability

In view of the international market intended for such devices, the TDA3562A has a dual-standard feature in that it can also work with NTSC signals. By a clever arrangement of mode-switching and 'dual-personality' of some lead-out pins, the PAL switch can be disabled and a tint control brought into operation by DC mode switching at pins 24 and 25 of the chip. If the NTSC subcarrier frequency differs from 4.43 MHz (as it does in the USA, but not in certain 'hybrid' signal formats used in the UK in connection with special VTR playback modes etc.) the transistor switches at the bottom of the diagram can be changed over to bring in a suitable reference crystal – the one shown is 7.16 MHz, twice the 3.58 MHz USA standard subcarrier frequency.

In combination with another chip, the device can also handle SECAM signals! Many colour decoder ICs can deal with PAL, SECAM and NTSC signals.

RGB video stages

It is in the post-demodulator stages that modern IC designs differ most from the traditional model of signal-processing practice. In Figure 6.6 only the blue channel is shown in its entirety – the R and G stages are identical. We left the Y signal at the RGB matrices, and it's time now to take up the story from there. The output from the matrix is of course in primary-colour form and it first encounters the *data switch*. For each primary-colour channel this forms a double-pole switch able to select either broadcast or locally-generated signals under the influence of the data blanking input on pin 9. When pin 9 is raised to logic level 1 the switch operates and locally-generated video data (digital signals, typically from a teletext or Prestel decoder, or a computer) is passed from the 'insertion' pins (12 14 and 16 for RGB respectively) to the primary-colour outputs. The data blanking input can operate at video rate if necessary, enabling superimposition of text or graphics on analogue-derived pictures; a good example is the 'mix' mode in text-equipped receivers. As with luminance and chrominance signal inputs to this IC, the data is AC-coupled in, then clamped within the chip.

Next comes the contrast control stage, and here again we have a DC control input feeding a voltage-controlled amplifier so that signals are not routed to and from the control panel, and facilitating remote cordless command of the set's functions. Because the voltage-controlled attenuator within the chip has an essentially anti-log control input/signal output relationship, and in view of the fact that linear control functions are easier to arrange from panel controls or remote-command decoders, the chip contains at pin 6 a lin/log converter – a similar device is present at pin 5, saturation control input. Because contrast is independently but simultaneously controlled in all three primary colour channels, matching between them must be very good; it can be seen that the same applies to the next (brightness control) stage, and that inserted RGB signals in the data blanking mode also come under the influence of contrast and brightness controls.

The brightness control works on a clamping basis, restoring the three black levels to a point set by the DC control voltage on pin 11. This is carried out during the '3L' period of black level insertion at the luminance amplifier, and described earlier.

Automatic grey-scale tracking

As we shall see in the next chapter, a colour picture tube contains three separate electron guns, one for each primary colour. For correct picture reproduction in black-and-white or colour it is essential that these three guns have exactly the same emission characteristics. If the cut-off points (beam-extinguish voltage) vary between them, for instance, the same signal applied to all three will give different results on each, so that we may find red-tinted low lights because the red gun has a higher (cathode) cut-off voltage than its companions. Alternatively, perhaps, the green gun will cut off at a lower cathode voltage than red or blue, leaving a magenta tint in place of dark grey in the display. These differences, initially due to manufacturing tolerances, can be easily compensated for by pre-set adjustment of black-level voltage for each individual gun. As the tube ages, however, differential drift of the gun characteristics takes place; this necessitates occasional resetting of the internal black-level controls during the life of the tube to maintain tint-free reproduction. For some years technology has been available to automatically adjust the black-level points for each individual gun. The TDA3562A has built-in auto black-level correction and we shall now examine the operation of this.

During TV lines 22, 23 and 24 and their companions 334, 335 and 336 on the twin-interlaced field, the scanning spot is at the top of the screen and out of sight, being just about to 'light up' to start describing the first active TV lines of the field. It is at this time that the auto-grey-scale compensation takes place. For the duration of these three lines the 3L insertion-pulse blanks all video information in the RGB channels; see the time-related waveforms of Figure 6.8. A 1-line duration clamp pulse is now inserted into each of the colour channels in turn; red, blue then green corresponding to L1, L2 and L3 in Figures 6.8 and 6.6.

This clamp pulse drives each gun of the picture tube just to the cut-off point and the beam current in the tube (nominally zero at black level, in practice a few microamps for measurement purposes) is measured and passed into the chip on its

Figure 6.8 *Time-related auto-grey-scale-shift and blanking waveforms for the TDA3562A IC. This diagram is centred on the field blanking interval*

76

pin 18. Thus pin 18 (row 8 in Figure 6.8) receives three pulses in quick succession, sequentially giving information on the cut-off points of the R, B and G guns of the tube in turn. These reference levels are gated to the three storage capacitors connected to pins 10, 20 and 21 respectively, and the charge stored in each of them is proportional to the cut-off point of the corresponding tube gun, which will vary with temperature and aging of the picture tube.

The charge on the storage capacitor is made to modify the amplitude of that gun's clamp pulse so that a separate feedback loop is maintained for each gun; with all three guns thus controlled, automatic grey-scale correction is achieved, and the need for a black-level adjustment preset in each channel is eliminated.

RGB output amplifiers

If the display device is a picture tube, either a shadow-mask device or three separate display tubes in a projection set-up, we need three matched-voltage amplifiers to raise the RGB signals to a level sufficient to drive the modulating electrodes, usually the cathodes. The bandwidth of these needs (if data and other video sources are to be handled) to be in excess of the 5 MHz or so of normal Y broadcast signals because rise and fall time (the time taken for a video signal to raise itself from black level to white, and vice-versa) is inversely proportional to bandwidth, as we saw in connection with Y/C registration and Y delay lines. Where data is being displayed the sharpness of the vertical edges of the characters depends on RGB bandwidth, the greater the better! As previously mentioned, the three amplifiers need to be closely matched in all respects as they separately handle components which are reblended in the display to form a composite and very critical image.

A typical video drive amplifier is shown in Figure 6.9. Here the video amplifier is Tr1 working in common-emitter mode with a gain of about 25. The video input signals to this stage are positive-going for 'white' and the polarity inversion in the stage renders negative-white output signals suitable for cathode drive of the picture tube. The difficulty in arranging the output amplifier is that it has to drive a wideband signal into the considerable capacitance of the tube cathode and its associated wiring; this calls for low output impedance. On transitions from black to white Tr1 is being driven on and its low impedance quickly discharges the stray capacitance to give a fast fall-time to the output waveform and a sharp leading vertical edge to the televised object, be it a data character or picture highlight.

When the transistor turns off, however (trailing vertical edge of picture object), the stray capacitance can only charge relatively slowly via the amplifier's load resistor R4 and because this RC time constant is long a trailing smear effect will be manifest on the picture. This is where Tr2 comes into operation, acting as a low-impedance emitter-follower with its base bias coming from R4. The shunt capacitance quickly charges via Tr2 and low-value R5, then, and crisp pictures are ensured by the low output impedance, and hence great bandwidth, of this video output stage. Two further identical amplifiers are needed in a practical TV, of course, and they share the emitter-potential reference diode D3 and work from a common HT supply of about 200 V.

Figure 6.9 *A wideband video amplifier for R, G or B channels. Circuit operation is explained in the text*

One other aspect of the RGB stages must be mentioned, and this is the necessity for drive adjustment. In Figure 6.9 R1 is made variable and acts as a 'contrast' control for the primary colour in question. The presence of similar control in the two other channels enables us to adjust the *drive* for reasons which will become plain in the next chapter. Where no auto-grey-scale correction is provided by the decoder chip, a further preset is required, this time to set the black-level for each gun. It effectively alters the standing voltage at the picture tube cathode to facilitate manual matching of the cut-off points of the three guns.

Decoder summary

In this chapter we have seen the 'chroma unscrambling' procedure in some detail, both from a theoretical point of view and in terms of modern hardware. A photograph of an RGB amplifier mounted on the picture-tube base panel is shown in Figure 6.10.

Figure 6.10 *Picture-tube base panels usually incorporate the final RGB amplifiers to reduce stray capacitance in the cathode wiring. Here the high-voltage output transistors are clearly identified by their metal heat-sinking tabs. The signals enter via the plug at top left*

7 TV display systems

A brief account of the operation of a monochrome picture tube was given in Chapter 1, and in this chapter we shall enlarge on that description to take in colour tubes and finally other types of colour display. A great deal of research and development has gone into the technology of colour picture tubes over the years, and direct-view types are now available with screen diagonals up to 89 cm. Improvements have been made in screen materials, gun technology and energy demand. The colour tube uses all the techniques of its simpler counterpart, and by extension and refinement of these is able to present a full colour picture on a single 'integrated' screen – no mean achievement! All direct-viewing colour tubes work on the shadowmask principle originally brought to fruition by Dr A. N Goldsmith and his research team in the American laboratories of RCA Ltd in 1950. The original device was rather clumsy and cumbersome by today's standards, but it embodied all the fundamental features of the current generation of colour tubes, and was amenable to mass production with all the cost advantages that could bring. The practical realisation of a relatively compact and decidedly cheap domestic colour TV receiver was (and is) dependent on the direct-viewing shadowmask tube concept.

The three-gun picture tube

Colour television picture tubes have to carry three different screen phosphors in order to produce three displays in perfect registration, one each in red, green and blue. If the face of such a tube (when operating) is examined under a magnifying glass the phosphor-dot formation might be seen, as shown in Figure 7.1. The phosphor make-up of the dots, of course, looks the same for each group of colour, and it is only when the phosphors are stimulated by the electron beams that the different colours are produced.

There thus exists a multiplicity of 'groups' of the three colour phosphors, and each group, detailed in Figure 7.1, is called a triad (*group of three*). In a typical tube there are almost 500 000 triads, together forming an interwoven pattern over the entire screen area.

The three guns are arranged in the tube neck so that the electron beam from one energises only the red phosphors, that from the second only the green phosphors

Figure 7.1 *The formation of the red-, green- and blue-glowing phosphor dots on the screen of the in-line shadowmask picture tube. The three-dot formation is called a triad*

and that from the third only the blue phosphors. For this reason the guns and beams are often referred to as red, green and blue. They are not coloured, of course! When a group of triads is caused to glow at proportioned colour intensities, the eye is deceived into perceiving that the three primary colours occur at the same point and so it discerns a spot or pixel of a colour corresponding to the mix of the three lights. When the mix is correctly proportioned for white light, the spot or pixel appears without colour as a 'shade' between white and black – i.e. a shade of grey.

Remembering our description in Chapter 2 of additive mixing by lights, it is easy to see how the spot or pixel can have any of the three primary colours red, green or blue; a complementary colour, yellow, cyan or magenta; or indeed any colour of intermediate value that can be produced by an admixture of the three primaries available, and falling within the 'television triangle' of the chromaticity diagram, Figure 2.3. This embraces a wide range of colours which make the colour TV display virtually subjectively equal to a good-quality colour photographic print.

The diagrams of Figure 7.2 give a basic impression of the working of the shadowmask three-gun picture tube. It is, in fact, three 'tubes' in one glass envelope and sharing the same screen. The electron-gun assembly has three cathodes which work quite independently to produce the three separate electron beams, each of a different strength according to the primary-colour signals by which they are being controlled.

Shadowmask

In addition to the components of the tube already mentioned, there is what is known as a shadowmask, from which the tube gets its name. In appearance this is something like a metal gauze pierced by a large number of holes. In fact there is one hole, or slot, for each triad and the purpose of the mask is to ensure that the phosphor dots of the different colours are hit only by the electron beams from the guns of corresponding colours. This will be made clearer shortly!

The shadowmask is located about 12mm behind the phosphor screen, and the three beams are made to converge on a single hole and cross over within it to

Figure 7.2 *These diagrams show how the three-gun tube can be regarded as 'three-tubes-in-one'*

diverge slightly beyond so that the red, green and blue beams strike their corresponding fluorescent colour phosphor dots.

Each dot is then activated by the electroluminescent process described in Chapter 1 to yield an amount of coloured light proportional to the intensity of the electron beam striking it. As already mentioned, the closeness of the dots and the integrating action of the human eye merges the three individual colours into a single spot whose colour and intensity are closely controlled by the video signal. A broad impression of the total TV illusion is given in Figure 7.3 which illustrates the simultaneous analysis, passage, reassembly and perception of a single pixel, of which many millions are sent every second!

The shadowmask is made of a thin sheet of metal identical in size, shape and contour to the glass faceplate of the tube. It is manufactured by a photochemical etching process similar to that used for the production of lithographic half-tone plates for printing. The positional and etch-timing accuracy has to be several orders higher for the shadowmask, however!

Arranging the phosphor dots

To produce the phosphor pattern on the inner surface of the glass faceplate a very ingenious method is used. The green glowing phosphor material is first sprayed onto the screen so that it covers the whole area, and then the shadowmask is mounted accurately in the faceplate. To 'fix' the green phosphor material on the

Figure 7.3 *Instantaneous reception of a colour element using a three-gun tube*

screen in the dot-pattern required, ultra-violet light is arranged to pass through the shadowmask holes at an angle corresponding to that of the electron beam from the green gun when the tube is correctly active. The effective origin of the green beam, so far as the shadowmask is concerned, is the *deflection centre*, a point near the physical centre of the scan coil's field. It is here that the ultra-violet light source is placed during the phosphor-fixing stage in manufacture.

When the green phosphor has been selectively hardened in the required dot pattern, the shadowmask is removed and the screen is washed to remove the unexposed phosphor material which was in the shadow of the mask. We now have a uniform pattern of green-glowing phosphor dot 'islands' in the 'sea' of the glass inner-screen surface.

This process is repeated separately for the red- and blue-glowing phosphor materials with the ultra-violet light in these cases positioned in the deflection-centre point corresponding to the colour being fixed. At the end of the operation, then, we have a regular pattern or mosaic of dot triads aligned with the shadowmask and with the incident angle of the electron beams from the right guns. The pattern is shown in Figure 7.1.

There are several refinements carried out on the basic dot-mosaic screen for the sake of good contrast ratio and high efficiency. We saw in Chapter 2 that the darkest part of the picture can only be as dark as the unenergised screen itself, and the phosphor materials used are pigmented and dark-coloured to absorb ambient light falling on the screen and prevent reflectance. Additionally the small space between the phosphor dots is filled with a similar light-absorbing black pigment; these artifices increase the contrast range. Like a monochrome tube the shadowmask type is aluminised and this highly reflective screen backing increases brightness by reflecting phosphor-light forward instead of wasting it within the internal bowl cavity. The aluminium layer also acts as final anode and forms a protective ion-barrier for the screen.

Tube construction

Before we go on to examine the operation of the shadowmask tube let us look at some of its physical characteristics, starting at the back. The heaters are similar in concept to those used in monochrome tubes, but their power consumption is lower, and a quick-heat facility (less than five seconds warm-up) is achieved by concentrating the heating element just behind the front surface of the cathode. The neck diameter varies with tube type, but 29mm is common in PIL tubes where a single 'gun' is used (still producing three beams of course!).

Deflection angles are generally 90° or 110°. Very small colour tubes used in portable colour TVs have deflection angles down to around 50°, while at the other end of the scale some of the Sony Trinitron types have a deflection angle of 114°, which represents about the limit of current technology. The conductive inner and outer coatings of the 'bowl' section of the tube are resistive rather than highly conductive in many colour tube designs; these form an EHT reservoir capacitor with the glass wall of the tube as dielectric. The resistive coatings absorb suddenly-released energy in the event of internal flashover within the tube, and harmlessly dissipate it over a large surface area. Coming now to the shadowmask itself, we have seen that correct alignment of the beams on their own phosphor dots is essential for correct operation and good purity of colour. The correct alignment of the shadowmask is crucial to this. Because only about 23 per cent of the mask area is 'holes' and the rest is 'wall', most of the energy in the electron beams is dissipated within the mask rather than at the phosphors, and the total current flowing in the mask can exceed 1 mA. With an EHT voltage in the region of 25kV the dissipation in the shadowmask can thus approach 30W, and if this is sustained for long appreciable heating of the mask will take place. The resulting expansion must be taken up in a defined and controlled way if buckling and distortion are to be avoided – these would upset beam landing and purity. The mask, then, is either fitted under tension (Trinitron vertical aperture-grille) or in a special frame incorporating bi-metal support clips to permit an axial movement of the mask to take up expansion while maintaining hole-triad registration. As we shall see, Trinitron tubes use a mask consisting of continuous vertical slots, and this is more vulnerable to heating, especially local heating at a point of high brightness in the display. To give lateral support to the vertical grille members in large-screen Trinitron tubes, three very fine horizontal platinum tie-bars are micro-welded across the grille at intervals.

Regarding the screen itself, the size quoted used to be the diagonal measurement from corner to corner of the *glass faceplate*, larger than the picture. Today, screen sizes are quoted in V (visible) sizes of the picture diagonal. This may vary from 10 cm to 89 cm, though most domestic sets are fitted with tubes whose diagonal ranges from 50 to 68 cm. The faceplate has two layers, the outer of which acts as an implosion guard in the event of sudden collapse of the glass envelope. These two front layers together are typically 2 cm thick in a 68 cm tube, and this largely accounts for the weight of a colour receiver! The atmospheric pressure on the envelope of an evacuated tube is very great.

The PIL tube

Having set the stage, as it were, we shall now look at the basic operation of the shadowmask tube, initially the PIL type. PIL stands for 'precision-in-line' and refers to the configuration of the emitting cathodes, and the electron beams as they travel down the tube. The single electron gun of the PIL tube is illustrated in Figure 7.4, where it can be seen that a common accelerating anode system and electron lens is shared by all three beams which travel through the gun assembly in very close proximity (5 mm) to each other. The three heated cathodes are in-line abreast at the left of the diagram, and their electron beams emerge through three small holes in the common control grid. They next encounter the first anode (A1) which accelerates them towards A2; the latter, in conjunction with the potentials on nearby A1 and A3 forms an electron lens for beam focusing. On emergence from the gun assembly the beams are on converging paths, as shown on the right of Figure 7.4.

In fact the beams' trajectory is such that they cross over in the shadowmask hole and diverge just beyond to strike their appointed phosphor dots - an idea of this is

Figure 7.4 *The internal arrangements of a PIL tube, viewed from above*

Figure 7.5 *Working principle of the shadowmask and phosphor screen. For each beam, all but the intended phosphor dot is in the shadow of the mask*

conveyed in Figure 7.5, where the centre beam (red in our example) is on the axis of the tube with the outer beams, green and blue, having an approach angle of around 1° with respect to red.

Purity

Clearly, the approach angle of the three beams to the mask must be spot-on to get the 'shadowing' effect right. If the effective origin of the beams does not exactly coincide with the position of the light source used for dot fixing during manufacture, one or more of the beams will spill over onto part of a neighbouring phosphor dot which will be the 'wrong' colour for that beam. If this happens to red, say, and it spills onto the blue dots the red gun will give rise to some blue light from the screen and a magenta 'stain' will mar what should be a pure red field. To prevent this a pair of two-pole ring magnets is provided around the tube neck in the gun area, and adjusted to achieve an on-axis beam path for good purity at screen centre. Secondly, the entire scan yoke is set axially (by sliding it along the tube neck) to establish coincidence between deflection centre and original dot-fixing light-source position. When this is correct, purity is established over the entire screen area for all three colours.

We have now achieved our aim, which is the establishment of three super-imposed images, one in each primary colour and independently controlled for brightness. The triple beam may be regarded as a tri-coloured light pencil, with each colour winking and flashing as the picture is scanned out by the common deflection fields to describe a full colour picture. There is another factor to be considered, however, and this is the necessity to *register* the three images so that they exactly overlay at all points on the display; this is called *convergence*.

Converging the beams

What Figure 7.5 does not convey is that each beam is larger than one shadow-mask hole, and in fact overlaps the slot edges. This means that not only must the beam approach angle be correct, but the spatial position of each beam must be adjustable to take up manufacturing tolerances and ensure that the three rasters are exactly superimposed; any error here will cause colour fringes on borders and outlines in the reproduced picture. What is required, then, is a means of inde-pendently shifting the beams in the tube neck, and this is arranged by further ring magnets around the tube neck to provide well-defined and adjustable magnetic fields embracing *individual* beam paths within the electron gun. These fields are set up by a combination of four-pole and six-pole ring magnets.

There is a *pair* of ring magnets to set up the four-pole field, and a further pair to provide the six-pole field. Two are provided in each case so that by relative- and co-adjustment the *strength and direction* of the resulting magnetic field can be manually set to give complete control of beam positioning. None of these fields affect the central beam which travels straight down the axis of the tube after the fashion of that of a monochrome tube – this implies that the rasters due to the outer beams (green and blue in our case) are made to conform to the position of the raster traced out by the centre (on-axis) beam. Let us look first at the effect of the four-pole field. It is shown in Figure 7.6 and can be seen to move the outer beams *differentially* in either vertical or horizontal planes, the direction of shift being determined by field direction, and the amount of shift by field strength. This permits us to register the blue and green images to give a cyan overlaid image, but this will not necessarily coincide with the red image, and this is where the six-pole field comes into play.

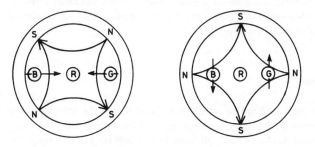

Figure 7.6 *Four-pole static convergence rings with field pattern*

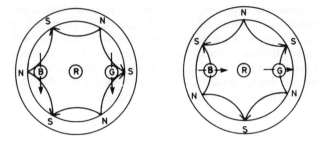

Figure 7.7 *Six-pole static convergence rings with field pattern*

The six-pole field is shown in Figure 7.7. Here the positions of the outer beams are moved together in the same direction, which may be vertical or horizontal in compliance with the orientation of the lines of flux. Again the flux strength (and hence the distance the beam moves) is controlled by the relative movement of the ring magnets, while the flux direction is varied by moving the six-pole rings *together*. Thus we can position our cyan raster exactly on top of the existing (and unvarying!) red raster.

On plain mixed-colour fields misregistration of the colours is not visible, so that convergence adjustment is carried out by observing the display of a crosshatch or dot-matrix pattern from a special signal generator.

Dynamic convergence

The convergence process described so far will only overlay the images in an area corresponding to a small circle at the centre of the screen. As the beams are deflected to left and right in the line-scanning process, the fact that the outer guns are off the tube axis means that the point of convergence will progressively fall short of the screen unless the latter is cylindrical with cylinder and deflection centres coincident as represented in Figure 7.8a. Because practical TV screens are much flatter than this, the three rasters become horizontally displaced (Figure 7.8b) and arrangements must be made to compensate for this. While a dynamic convergence system could be devised along the lines of the static magnets already described, but with line-rate flux variations (i.e. a strap-on convergence yoke fed from the line timebase), a much neater system has been evolved, based on the concept of a special deflection yoke whose magnetic field is tailored to achieve correct dynamic convergence.

PIL deflection yoke

The degree of deflection suffered by an electron beam under the influence of a magnetic field depends on the strength of that field, i.e. how many lines of force it cuts. In a homogenous (uniform) field, all three beams will be 'bent' through the same angle as in our Figure 7.8a. Because the beams are horizontally separated

Figure 7.8 *The causes of misconvergence: (a) shows how the beams diverge in areas away from screen centre, while (b) indicates the resulting misregistration*

by a few millimetres on their journey through the deflection field, however, we can arrange for them to encounter different flux levels by carefully 'grading' the magnetic deflection field across the tube-neck diameter; see Figure 7.9. Here the lines of force are widely separated in the area of the tube axis, and become progressively closer towards the tube's outer walls. This means that each of the outer beams will, during horizontal deflection, be turned into an area of stronger or weaker flux than its fellows, and so suffer a greater or lesser deflecting force. Provided the flux gradient across the tube cross-section is just right, complete convergence of the rasters can be achieved.

This effect is clarified by the beam trajectories in Figure 7.9. This is drawn from above and taking the case (solid line) of deflection to the right, we can see that the central red beam passes through an angle determined largely by the magnetic flux at the centre of the field to strike the fluorescent screen at point A. The blue beam, if it were similarly deflected, would cross the red beam some way short of the shadowmask and land at imaginary point B causing misconvergence. However, the blue beam can be seen to be (at the deflection point) turning into a weaker field, causing it to be deflected through a lesser angle than red and 'aiming' it precisely at screen point A. Now consider the green beam. In a uniform deflection field it would cross the red beam too early and diverge to strike the screen around point C. This time, then, we need to increase the green beam's deflection angle to achieve screen registration; as the green beam passes through the deflection field it is turned into a stronger deflecting field to achieve this. For line-scanning deflection to the left the opposite happens; in all cases the beam on the inside of the deflection-centre 'bend' is bent through a greater angle than the norm, and the beam on the outside is bent through a lesser angle.

What happens during vertical deflection? Again the yoke-to-screen beam-path gets progressively longer as we move upwards or downwards from screen centre,

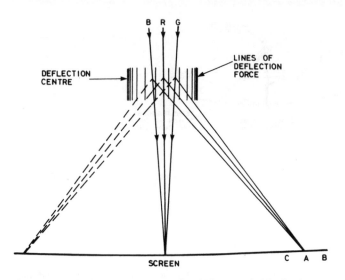

Figure 7.9 *Beam trajectories in the deflection field. The latter is non-uniform to achieve correct convergence, as explained in the text*

so that *horizontal* displacement of the three images will take place with increasing severity towards screen top and bottom. Again an astigmatic deflection field will correct this in similar fashion to that described for horizontal dynamic convergence. The vertical deflection field is barrel-shaped (Figure 7.10) so that the required horizontal lines of force have an increasing vertical component away from the tube axis. This vertical component has the effect of reducing the *horizontal* divergence of the outer beams to achieve horizontal register at screen top and bottom.

Thus correct convergence is maintained all over the screen area by a precisely-tailored deflection yoke. The positioning of the yoke along the tube axis is

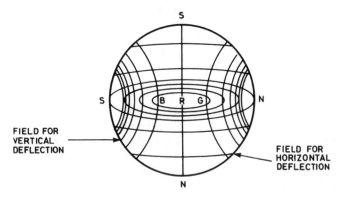

Figure 7.10 *The astigmatic deflection fields required for a self-converging in-line picture tube. The field shapes shown are achieved by precision construction of the deflection yoke*

dictated, as we have seen, by the requirements of purity setting, and the gun end of the yoke is concentric with the tube neck due to their common diameter. To take up manufacturing tolerances, however, the flared front end of the scan yoke assembly may be tilted in both vertical and horizontal planes to 'trim' the relative trajectories of the electron beams and thus form a final adjustment of screen-edge convergence; this adjustment is carried out during manufacture or as a servicing routine, after which the yoke is wedged and sealed in position.

The necessary very precise field shaping is achieved in the PIL tube by a toroidal deflection yoke in which the positioning and distribution of the individual turns of wire are defined by notches cast in the supporting frame. The fields are further modified by magnetic 'shunts' and 'enhancers' fitted inside the tube or the scan yoke which respectively disperse or concentrate local magnetic flux to create exactly the field shape required for self-convergence.

20AX system

An important milestone in in-line shadowmask tube technology was the Mullard/Philips 20AX system. The screen and shadowmask of 20AX was similar in concept and construction to that already described for the PIL tube, and while PIL types come in various sizes and had deflection angles of 90° and 110°, 20AX was confined to 56 and 66 cm screen sizes and 110° deflection. Instead of a single gun shooting three beams, this tube contains three separate guns, each very similar to that of a monochrome tube. The centre gun (green) is on the tube axis and the outer guns, for red and blue, are tilted inwards so that all three beams cross over within the shadowmask holes at screen centre. The construction of the 20AX gun assembly is shown in Figure 7.11.

The ring magnet system is very similar to that already described for PIL, with four- and six-pole magnet rings to set static convergence and a two-pole pair for

Beams converge at slot-mask

Figure 7.11 *The 20AX gun assembly. Three separate electron guns send their beams on a converging path*

purity adjustment. 20AX also uses an additional two-pole ring pair for geometic raster correction (symmetry), which is adjusted to eliminate bowing of the horizontal centre line.

The yoke of the 20AX system achieves the astigmatic field pattern described earlier for PIL tubes by a saddle-shaped winding pattern whose appearance is broadly similar to that of an ordinary monochrome deflection yoke. However, the wires are grouped into 'streams' to create the right field pattern, and the tube-internal magnetic components are not present in a 20AX type. Very close-tolerance manufacturing techniques in both yoke and tube eliminate the need for yoke manipulation in dynamic convergence setting. The yoke remains firmly on the axis of the tube neck, and any 'trimming' adjustments for dynamic convergence are catered for by a simple four-pole convergence correction coil associated with the deflection yoke. This carries sawtooth currents at line and field rate, whose amplitude and direction are controlled by two or three preset controls or links; further convergence correction is made by differential adjustment of the currents in the symmetrical halves of each section (vertical and horizontal) of the deflection coils themselves.

30AX

Based on the technology of the 20AX system and the manufacturing experience gained, further developments led to the release of the 30AX display system, in which all the tube-neck adjustments were eliminated. These tubes were very similar in concept to the 20AX type, sharing the triple-gun and thick-neck characteristics and retaining 110° wide-angle deflection, but having a wider range of sizes, e.g. 50, 56 and 66cm.

The 30AX gun assembly is longer than that of the 20AX because it has to accommodate a weaker electron-lens system – this gives improved overall focus. The 30AX has no neck-mounted magnetic rings; their place is taken by a small magnetic wire ring at the top of the gun assembly, into which is printed, at the time of manufacture, a composite magnetic field to fully correct convergence, purity and geometry. For a given tube size 30AX deflection yokes are fully interchangeable.

45AX

A later Mullard/Philips design, designated 45AX, had a *unitised* gun similar to those described above in the PIL section. Here all electrodes downstream of the cathode are 'commoned' for all three beams, whose trajectories are governed together by the deflection system; and individually (for static convergence, geometry and purity purposes) by the single integral magnetic ring system pioneered in the 30AX. The main development here, however, is in the mounting of the shadowmask on a light diaphragm with a special suspension system to accommodate expansion of the mask with electron-beam heating effects. This greatly reduces the risk of purity errors.

FST

Current designs of TV receivers use FST (flat, square tube) designs in which the faceplate has a radius of curvature about twice that of a conventional tube, with straighter sides and squarer corners. The effect is to reduce ambient light reflections and give a more pleasing appearance to the set.

The Trinitron

The third group of in-line tube designs is represented by the Trinitron type which in fact was the first in the field, having been introduced by Sony in 1969. The main difference between this and the types so far described is in the form of the shadowmask. Instead of the discrete phosphor-dot triads of the conventional tubes, the screen of the Trinitron is composed of several hundred vertical stripes of red-, green- and blue-emitting phosphors. The mask consists of a metal aperture grille which has one continuous vertical slot for each three (R, G, B) phosphor stripes. The general principles of the design are illustrated in Figure 7.12. It can be seen that (unusually amongst colour tubes) the beams cross over within the single gun assembly before crossing again at the aperture-grille slot.

The RGB beam crossover within the focus field confers the advantage on this design of having all three beams pass through the centre of a relatively large electron lens. The analogy to light beams in optical lenses is obvious, and focal aberrations in the 'outer' beams are thus minimised. A further 'electron-optic' function is carried out further downstream by the deflector-plate system shown in the middle of Figure 7.12. These plates act as a prism due to a PD (potential difference) between them of a few hundred volts. By adjusting this PD the outer beams (red and blue) can be shifted laterally in opposite directions to converge them with the central green beam. The static convergence control, then, is associated with the EHT voltage source and connects to the prism plates via a coaxial EHT cable.

In other respects the Trinitron is similar to the PIL tube already described. Vertical static convergence is catered for by adjustable neck-mounted perma-

Figure 7.12 *The main features of the Trinitron single-gun, three-beam aperture-grille picture tube*

nent-magnet systems, and such dynamic convergence trimming as is required is carried out by tilting the scan yoke. As in 20AX, a simple four-pole dynamic convergence coil is present and if necessary (depending on manufacturing tolerances) energised by a line-rate (and in large screen types, a field-rate) adjustable sawtooth current derived from the timebase section.

Two advantages which the Trinitron offers over the opposition are the increased mask 'transparency', leading to higher picture brightness, and the enhanced vertical resolution, both due to the absence of a vertical matrix component in the shadowmask design.

Degaussing

Any external magnetic field will upset the beam path in a picture tube as a result of which beam landing accuracy – and hence purity – will be affected. Even the vertical component of the Earth's magnetic field can have an influence (the horizontal component cannot upset operation of an in-line gun tube) and nearby magnetic objects such as loudspeakers, radiators and steel joists can give rise to sufficient field strength to spoil the display.

To prevent these effects, the flare of the tube has a built-in magnetic shield, and to neutralise any magnetic field it may itself collect, a built-in degaussing system is provided in the receiver. It takes the form of a pair of coils embracing the tube flare which are briefly energised at switch-on by a decaying burst of 50 Hz mains power. This comes from a double PTC (positive-temperature-coefficient) device whose resistance rapidly increases with the heating effect of the mains current so that the degauss energy decays to zero after two or three seconds, and remains so due to the presence of a 'bleed' resistor which keeps the PTC combination warm. To prevent pickup and distribution of line scan-yoke energy the degauss coils are bypassed at TV line frequency by a suitable capacitor.

Geometric distortion

In a monochrome picture tube, especially a wide-angle type, the flatness of the screen leads to a form of raster distortion known as pincushion, in which the raster borders are bowed in. This is corrected by permanent magnets which neutralise the effect by applying an equal and opposite pre-distortion of the scanning field. Colour tubes have similarly flat faces and so the same inherent defect is present, but any permanent-magnetic correction system is not practical for reasons which should now be obvious! Instead the pincushion correction is carried out dynamically, as it were, by amplitude-modulating the line-scanning current with a field rate *parabolic* wave-form, and by superimposing a line-rate sawtooth wave on the field-scanning current.

The degree of pincushion distortion which has to be catered for depends on the type and vintage of the tube in question; early PIL types needed considerable correction on both axes, whereas later designs eliminated the need for vertical correction and reduced the horizontal requirement. Current tubes, by further refinement of the yoke design, eliminate the need for pincushion correction alto-gether, and are called 'pin-free'. 20AX and 30AX require no vertical pincushion correction and 12 and 8 per cent respectively in the horizontal direction. This latter is arranged by a *diode modulator* in the line-scan stage which alters its tuning, and hence scan amplitude, on a field-rate basis.

Grey-scale tracking

A very important requirement is for the beam current versus cathode voltage characteristic of the three guns to match over the full range of brightness. Since it is impossible for guns to be manufactured with these parameters exactly match-ing (and to expect them to wear at equal rates), the tube has to be equipped with various adjustments to provide accurate *grey-scale tracking* as it is called.

We have seen that white light is produced when the mix of red, green and blue lights is correctly proportioned. This must occur at all drive levels from black to full brightness for each gun, for a drift out of alignment as drive is changed would alter the relative intensities of the three colours and thus change the proportions of the mixture, which would result in the white (or grey) display becoming coloured. The effect on a colour scene would be to alter the hue or saturation with changing luminance level. The black cut-off points are matched by varying the clamp potential on the individual cathodes by manual or automatic means, as we saw in the previous chapter. To match the brightness of the primary colours at high drive levels, the 'slope' of the video amplifier signal is adjusted by drive controls ('primary-colour contrast pre-sets') with which the highlights of a monochrome display are set to eliminate any coloured tint.

Focusing

We have talked much in this chapter of converging the beams for the purposes of picture registration, and this may be regarded as a focusing of the three beams inasmuch as we aim to get them to cross over at the shadowmask holes. Each individual beam may be likened to that in a monochrome tube, however, and the necessity to focus the beam itself is as important as in a black-and-white set; if good focus is not achieved, the colour image, while remaining converged, will have poor definition and render a blurred picture.

Each electron beam consists of a multitude of negative 'charge packets' – electrons – jostling and falling over each other in their headlong dash towards the highly-positive final anode. The fact that they are all *negative* charges makes them tend to repel one another, 'spreading' the beam. To correct this they pass through an electrostatic lens system (Figures 7.4, 7.11 and 7.12) in which the *outer* electrons in *each* beam are sent on an inward trajectory such that all the electrons in the beam come to a sharp focus at the phosphor screen. Adjustment of the focal length of this electron lens is carried out by varying the voltage on the focus anode(s) – usually A2; A3 in Trinitron types – by means of a potentiometer associated with the diode-split EHT transformer. Typically, the potential required is about 18 per cent of final anode voltage in the conventional *bi-potential* tube type, and around 27 per cent of final anode voltage in the more modern *high bi-potential* tubes. With a 25 kV EHT voltage in both cases, this calls for around 4.5 kV and 6.8 kV respectively, with a focus control range of about ±15 per cent. This potential is applied to one of the CRT base pins, and to avoid insulation and sparking problems the adjacent pins are widely spaced from it.

Beam current limiting

In our earlier description of the shadowmask, mention was made of the heating effect of the beam current flowing in the mask, and its special expansion arrangements to compensate. This works up to a limit, but at around the 1.2 mA mark (depending on screen size) the mask is at the end of its tether, metaphorically and literally! Beyond this point, purity suffers and convergence and focus become degraded. The *average* beam current in the tube can be measured indirectly by a monitor in the 'cold' side of the EHT supply, or directly at the cathodes. When the beam current reaches the upper limit, tube drive is automaticallly pulled back to prevent picture distortion. This 'pull-back' effect is carried out via the decoder (or luminance) IC by reducing the voltage available to the contrast, brightness and colour control-potential inputs.

Colour-tube performance

We have seen that there is one hole in the shadowmask per dot-triad, and because a TV pixel can be no smaller than a single triad, the definition capability of the tube is determined by the number of triads on the screen surface, assuming all other conditions of scanning, focus and convergence are optimum. In general the mask and dot structure are the limiting factors for resolution overall in a receiver, especially in the smaller screen sizes where the mask/hole matrix is not reduced in size in proportion to the screen area. Current medium- and large-screen direct viewing tubes are adequate for 625/50 colour reception, where the limitations on broadcast bandwidth offer a picture definition which can just be resolved by these tubes.

For higher-definition pictures and particularly high-resolution data and graphics displays, fine-matrix tubes are appearing with more triads per square cm, leading to an inherently better resolution capability. This puts further demands on the convergence system, which in any tube should ideally converge all three rasters to within one pixel over virtually all the screen area.

Another changing feature is the shape of the viewing screen. The new generation tubes have squarer corners and straighter sides than conventional types, and conform to the ideal 'rectangle' shape with an aspect ratio (width-to-height) of 4.3. This is the logical conclusion of the trend over the years, starting with a circular tube surrounded by a quasi-rectangular mask. For the PAL-Plus and Hi-Def systems described later in this book, 16:9 aspect ratio tubes are made.

Projection TV

For a large picture the direct-viewing system we have been describing is not practical for many reasons, foremost of which are weight, bulk and the effect of atmospheric pressure on a large evacuated glass envelope! Until a practical large flat-screen device is developed (no easy matter, as we shall see) the only alternative to direct-viewing CRT (cathode ray tube) displays is some form of projection system. The two basic approaches to this are to project the coloured image from a conventional-format tube onto a large screen by means of a lens; or to project separate RGB images onto a common viewing screen, rather like the set-up we used in Chapter 3 to illustrate the basic concept of colour TV and light mixing. The former approach has been used in domestic applications, but its light output is pitifully low since the limited CRT phosphor-light becomes spread over a large area. RGB systems are more complex and expensive, but are capable of much brighter picture reproduction for reasons which will become clear, and it is on these that we shall concentrate now.

In a projection system, many configurations are possible, and much depends on the required screen size and viewing area. To make a (relatively) compact set, the light path is often folded by an intermediate mirror and the image formed on an integral screen. This system may use *back-projection*, where the picture is built up on the rear of a ground-glass screen which (for domestic and small-audience commercial use) is fitted into a free-standing cabinet which thus resembles a huge conventional console TV set. An alternative and more common approach is the front-projection system where projector and viewers are on the same side of a reflective screen. This offers the great advantage of being able to give the screen a 'gain' factor by building a directional property into it – the concave screen throws a lot of light over a small viewing area and a useful increase in brightness is realised in the limited viewing area. Off-beam the brightness is very low indeed. A light 'gain' of three to five times is possible with a directional screen.

In the sort of RGB projection system we are discussing here, the light sources are three separate picture tubes with different phosphor screens, one each of R-, G- and B-emitting – corresponding to the three TV primary colours. The chief requirement is high brightness and with no shadowmask to intercept beam energy, and a high accelerating (EHT) voltage, the limiting factor for brightness is the heating effect of the beam dissipation on the glass faceplate. Very expensive projection tubes are made with sapphire faceplates to withstand high temperatures, but a more homely arrangement is the use of a forced-air cooling system for three conventional small cathode ray tubes. These may be mounted in-line abreast like the guns of a modern picture tube, or in delta formation analogous to the gun arrangement in early shadowmask picture tubes.

The next operation is to gather as much light as possible from each tube and project it onto the viewing screen, along a straight or folded light path. Either conventional lenses or a *Schmidt* optical system can be used; the basis of the Schmidt system is a highly-polished spherical mirror which gathers the scattering light rays from the tube and concentrates them into a parallel beam, the reverse process to that used in astronomical telescopes, or indeed the satellite receiving dish to be described in Chapter 13. The parallel beam passes through a corrector

PHOSPHOR COATED
INTERNAL SCREEN
(RED BLUE OR GREEN)

IMAGE PROJECTED
TO SCREEN

ELECTRON
BEAM

CORRECTOR
LENS

ELECTRON
GUN

EVACUATED GLASS
ENVELOPE

SPHERICAL PROJECTION
MIRROR

POSITION OF
DEFLECTION YOKE

Figure 7.13 *An internal-Schmidt projection tube (LightGuide) by Advent Corp of USA*

lens on its way to the viewing screen. To prevent contamination of the highly-polished reflector and to ensure perfect and permanent positioning, the entire Schmidt system is built into the tube envelope in some designs – Figure 7.13 represents the LightGuide tube by Advent Corp of the USA. The internal screen is about 9mm diagonal, and made of aluminium, coated with R, G or B phosphor.

Whether simple lenses or Schmidt units are used, the fact that the three projection lights are not coincident in position means that the light beams strike the viewing screen at different angles and, as with our shadowmask picture tubes, this causes misregistration of the three primary colour images. To correct this, a convergence system is incorporated in the deflection system of two or more tubes to introduce an equal-and-opposite distortion to each individual raster, and these are finely adjusted to overlay the three images on the viewing screen. Where the screen (or the incident light beams) is at an angle to 'normal', keystone distortion will also be present, and this is corrected in the same way.

Light valves

In all the systems discussed above for TV display, the basic source of illumination is phosphor-light, and for very bright or very large displays this technology is pushed to the limit. If we were able to emulate ciné practice, and use the picture information to *modulate* a separate, external light source on its way to the viewing screen, overall illumination would be limited only by the brightness of the 'lighthouse.' The first system of this sort used the Eidophor principle, wherein the TV raster is traced out on a spherical mirror in an evacuated glass envelope. The face of the mirror is coated with a special oil whose surface is deformed by the electron beam. The TV lines are formed as grooves in the oil film, and the depth and wall-angle of these grooves depends on beam intensity – picture brightness. A carefully-formed and precisely-angled light is focused onto the oil film which then passes or blocks the light so that the spherical mirror's reflectance is controlled by the TV signal. Emerging light is focused on the viewing screen by a lens, and the whole shooting-match is triplicated for the three primaries of colour.

LCD displays

The most common alternative to CRT technology for producing TV pictures is the LCD (liquid crystal display) system, consisting of a matrix of crystal cells. The basic difference between a picture tube and an LCD panel is that the former generates point-sources of coloured light while the LCD controls, at each picture element, the amount of light passing through it. The basic property of a liquid crystal is that its molecules are arranged in the form of thin bars or flat plates, and of many configurations; the *nematic* type is used for TV and has long, thin molecules, all aligned in the same direction. During manufacture the molecular structure is twisted through 90° (see Figure 7.14) so that the polarisation of light passing through the molecules is similarly twisted. The degree of twist of the molecules can be controlled by applying an electrical field at right-angles to their axes, as shown in Figure 7.15a. An LCD element has a ten-micron (1 micron = 10^{-6} metre) liquid-crystal layer sandwiched between two transparent electrodes, and when the voltage applied to the electrodes is varied, the rotational polarisation of the

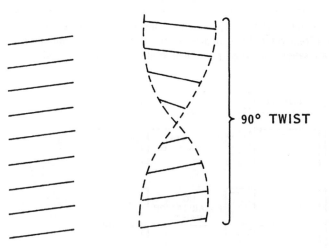

(a)   (b)

Figure 7.14 *Molecular structure of a nematic liquid crystal: (a) natural state; (b) twisted for use in LCD display panel*

(a)

(b)

Figure 7.15 *(a) As the potentiometer is turned up, progressively more voltage is applied to the LCD cell, increasing its molecular twist. (b) Twist is in proportion to applied voltage*

molecules in the crystal can be varied between zero and 90°; the effect is illustrated in Figure 7.15b.

The basic action of an LCD is shown in Figure 7.16. The polarisation of natural light is random: its photons vibrate in all directions. A polarising filter can be used to select light which is polarised in only one plane. The LCD element is sandwiched between two polarising filters with opposite characteristics, one passing vertically-polarised light and the other horizontally-polarised light. The 'transparency' of the assembly depends, then, on the crystal twist and hence the voltage applied. As the drive voltage is increased, the light passing through the crystal becomes progressively more at odds with the filter on the output side until a point is reached where all the incoming light is blocked and the screen is black. In some types of LCD the two polarising filters have identical characteristics so that the pixel is black for zero drive voltage and white (transparent) when the drive voltage is maximum. The voltage/opacity characteristic of the cell is almost linear (Figure 17.15b) and the drive voltage can be applied in either direction; in fact, to prevent ionisation and consequent deterioration, AC drive is used.

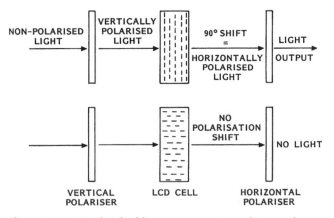

Figure 7.16 *Sandwiched between opposing polarisers, the crystal layer can be made to act as a light cell*

In a practical display panel, between 100 000 and 400 000 cells are arranged in a rectangular matrix, each cell being controlled by its own built-in thin-film field-effect transistor (TFT). By individually controlling the conduction of each TFT on the basis of a rows-and-columns drive signal, each picture element can be opened, closed or driven to any point between the two extremes to build up a pattern of light and shade, corresponding to a black and white TV picture whose definition is governed by the number of elements in the matrix. The drive circuit is integral with the LCD display signal, and consists of a series of shift registers, sample-and-hold stages and output buffers. When a TFT is switched on, once per TV frame, the display data for the pixel it represents is stored as an electric charge in the capacitor formed by the LCD cell. The charge remains, and the cell's light transmittance is held at the same level, until the pixel is updated when the next frame is scanned. This gives the best possible picture contrast and brightness.

Colour LCD displays

The panel described above gives a monochrome picture, not very popular with TV viewers! To produce a colour display each individual cell is given its own colour filter, producing an RGB matrix like that shown in Figure 7.17, with the RGB video drive signals routed to the appropriate LCD cells. Because we now have three cells per pixel, the definition capability is reduced – it's overcome by using more and smaller pixels, though the complexity of the drive circuits is increased in proportion.

LCD back-lighting

As we've seen, the LCD panel operates basically as a multi-point light valve, and needs a light source to produce a picture. It's generally provided in the form of

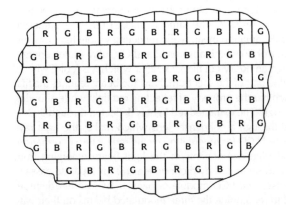

Figure 7.17 *An offset RGB matrix gives bèst possible definition in a colour LCD display panel. The same applies to CCD image sensors, which generally use* complementary *colour-matrix overlays*

one or more fluorescent tubes with reflector and diffuser to give bright and even illumination over the entire screen surface. In pocket and portable TV sets and LCD camcorder viewfinders, the fluorescent lamp is very tiny.

LCD projectors

As we have seen, the brightness produced by any sort of cathode ray tube is limited by the energy that can be imparted to its electron beam and dissipated at its screen. When (as with a three-tube projector) the available light has to be spread over a large picture area the result, even with a directional viewing screen, may be barely adequate. Because an LCD panel works on the principle of a light-valve there are no such constraints when it is made the active element in a TV projector: a very bright lamp can be used as the light source, its output being intensity- and colour-modulated by the LCD matrix between the lamp and the screen, which latter can now be made flat, cinema style, needing no 'gain' by virtue of directional properties – indeed a white wall can be satisfactorily used as a reflective screen.

Some LCD projectors use a single RGB panel of the sort described above, with a powerful backlight and a lens system, but it's difficult to achieve sufficient definition in a large picture projected from an LCD panel in which three cells are required for each pixel. Commercially-produced LCD projectors for home and industrial use have three separate LCD panels; the beam coming from the backlight is separated into its R, G and B components, separately modulated then recombined into a single beam for projection.

A typical three-LCD panel projector offers a horizontal picture resolution of 300–450 lines, using three separate (RGB) panels each with a diagonal measurement of about 8 cm, and each containing over 100 000 cells/pixels. Thus the display carries 300 000 or more transistors. A 150 W metal-halide lamp is provided as light source, capable of giving a contrast ratio of 100:1 via an f4.5 projection lens with a 'zoom' facility of about 2:1. The internal arrangement of the projector is shown in Figure 7.18. The lamp is backed up by a parabolic reflector (1) which absorbs IR radiation (heat) while reflecting visible light forward for reflection from a cold mirror (2) which absorbs IR and UV (ultra-violet) energy. Next the light passes through an IR/UV cut filter, (3), after which it is 'cold' with its energy confined to the visible part of the spectrum. Now the beam reaches the first (red) dichroic lens, (4) which has the property of reflecting red light while allowing blue and green to pass through. The extracted red light is deflected through 90° by a reflector (5), and passed through a condenser lens which makes the light rays parallel on their way to the red LCD for modulation by the red picture information. Meanwhile, the green and blue components of the light beam emerging from mirror (4) are intercepted and separated by dichoric lens (6) for separate passage to their own condenser lenses and LCD modulator panels. Two further dichroic lenses (7) and (8) are used to recombine the three modulated beams on their way to the projection lens and hence the screen.

The running temperature of the LCD panels is held below 50°C by a forced-fan ventilation system which sucks in air at the back of the projector housing and

Figure 7.18 *Principle of operation of a three-panel LCD projector for domestic use*

blows it out of the top. Internal temperature is monitored by sensors hooked to a control system capable of shutting down the lamp if necessary.

Study of Figure 7.18 shows that the image produced by the R-LCD is not reflected at all on its way to the screen, while that from the B-LCD is reflected once and from the G-LCD, twice. The result would be that the blue component of the picture would be horizontally reversed compared to the G and R images. It's compensated for by electrically scanning the B-LCD in the opposite direction to the other two.

Projector lamp

The metal-halide lamp used as light source is very powerful and efficient. It consists of a mercury-arc lamp with its electrodes 1-2 mm apart; initially an arc is struck between them by application of a 12kV pulse. The resulting discharge vaporises a drop of liquid mercury which radiates an intense glow, and establishes, when warm, a low resistance path between the electrodes. At this point an AC feed voltage of about 80 V is sufficient to maintain conduction and light emission: it's regulated by a control circuit which holds the lamp dissipation at about 150 W.

It's important that the spectral emission of the lamp is fairly evenly divided across the visible spectrum, and a standard mercury-arc lamp has a very uneven distribution, with its major component in the blue and UV region. To overcome this, various metal halides, e.g. caesium, are added to the mercury in the quartz glass envelope, which also contains an atmosphere of argon gas. The result is an even spectral distribution from 380-760 nm. The lamp's operating life is about 1000 hours, monitored by the system-control section of the projector.

8 The TV receiver

A colour TV set can be regarded as a monochrome receiver with refinements, to which has been added the colour display device and a decoder. Figure 8.1 shows the block diagram of a colour receiver with the 'extra' items required for colour in heavy line.

The colour display device is shown in heavy line because this is specific to a colour receiver, but as we have seen, a colour set can display very satisfactory monochrome pictures, whether or not the transmission is in colour. Thus the main and essential item in the colour set, apart from the display device itself, is the decoder, which has been fully dealt with in Chapter 6. In most other respects, monochrome and colour receivers are similar or identical – each contains a tuner with some means of viewer control; an IF amplifier with carefully-shaped frequency response; a vision detector to produce video and sound signals; an 'intercarrier' audio IF and detector stage; one or more audio amplifiers and loudspeakers; a sync-pulse separator; line and field timebase; and a power supply unit. The degree of sophistication of many of these circuit blocks needs to be greater for colour TV, and much depends on the price of the model in question, especially in such areas as remote-control, sound quality and interfacing ability. Many colour TV

Figure 8.1 *Block diagram of colour receiver. The heavy-line blocks represent the extra stages required for colour*

receivers are fitted with Teletext facilities. In recent years the production volume of black-and-white sets has been very low, consisting in the main of portable, small-screen types.

The function and operation of the various 'common' circuit sections are outlined in this chapter. More details are given in other reference books, e.g. *TV and Video Engineers Pocket Book* by the same author.

The tuner

The tuner's job is to amplify the weak signals picked up by the aerial, to select or 'tune' the required one, then to amplify and translate it, complete with all the information that it carries, to a lower (intermediate) frequency called the 'IF'. It will be recalled that the aerial itself is a tuned circuit, so the aerial and tuner should be regarded as a 'matched pair', a point perhaps which should be borne in mind by the aerial-erecting fraternity!

It should be remembered that the selected signals are in a specific channel and consists of both the modulated sound and vision carriers: Reasonable balance between all components of the transmitted signal on any one channel is required, though FM sound limiting and the ACC function make the set reasonably tolerant of 'tilt' in the response of the aerial or tuner section. Most of the gain and selectivity of a TV receiver is designed into the IF amplifier, and the main requirements of the tuner are a low inherent noise figure and an ability to suppress the 'image' frequency, which the IF amplifier is quite unable to reject. Normally the local oscillator (these tuners are, of course, superhet devices) runs 'high' with respect to incoming frequency, and the IF output from the tuner is the *difference* between the two, normally 39.5 MHz for vision. There are two input frequencies which will give rise to this 'beat' frequency, one 39.5 MHz above the oscillator frequency, one 39.5 MHz below. The former is the so-called 'image' and any signals arriving at the tuner on this wavelength must be strongly rejected.

The tuned circuits in the tuner unit must keep in step with each other as different frequencies are selected, and this calls for good matching in manufacture. The variable tuning is carried out by means of fixed Lecher bars (inductors), in conjunction with varicap diodes which we met in our discussion on the colour decoders. Tuning control, then, is carried out by varying the DC potential applied to them. The control voltage source may be as simple as a single potentiometer or as complex as a frequency-synthesis, self-seeking ensemble. Perhaps these last terms deserve a little more explanation! A self-seeking system, when initiated, sweeps up the TV transmission band(s) by itself, stopping each time it encounters a station for committing (manually or automatically) to its solid-state memory. Frequency-synthesis tuning offers a self-seek and memory facility along with a 'direct addressing' feature in which a required channel *number* (21 to 68 for UHF) can be requested by the viewer and automatically tuned. This involves a very stable crystal oscillator in a PLL embracing the tuner's local oscillator. The PLL includes a *programmable divider* to set the ratio between crystal and local oscillator and this division ratio governs the tuning point of the TV. All this is achieved by special digital ICs.

IF amplifier

Looking again at Figure 8.1, the level of IF signals (both sound and vision) from the tuner are raised by the IF amplifier. The response characteristic is carefully tailored to selectively amplify sound, chroma and vision signals, and reject out-of-band products, by a SAW (surface-acoustic-wave) filter, which may almost be regarded as a cross between a crystal and a glass delay line! With selectivity provided by this filter, and gain by the following amplifier, the IF signal is brought to a condition suitable for demodulation, or detection.

Vision AGC

Automatic gain control is applied to both the tuner RF amplifier and the IF amplifier. The vision signal is *sampled* (usually at line rate) to measure its amplitude, and the sample is produced in DC form as an *error* voltage or current whose level is proportional to signal strength. This is fed back to an early stage in a way that varies its gain, thus setting up a feedback loop which maintains a constant output level from the vision detector.

While the *level* of signal can be held constant by AGC over a tremendous range of RF input levels, the *noise* (snow/grain effect) is proportional to received signal strength. Below about 300 mV (for modern UHF tuners) this noise becomes obtrusive, and to achieve the lowest possible noise level in weak signals the gain of the tuner is held at maximum unless the signal becomes so strong (2–3 mV) that overloading of the tuner could occur, when its gain is automatically reduced. This is called *delayed* AGC.

Vision detection

The IF signal is demodulated within an IC by a synchronous detector working on a sampling system similar in essence to that employed for colour demodulation in a decoder. In place of the local oscillator we have a 'tank' circuit which is energised by the incoming IF signal itself, and critically tuned for linear demodulation. This is followed by a filter to remove residual IF carrier components. The vision signal is now ready for application to the decoder block as described in Chapter 6.

Intercarrier sound

At the vision detector input both vision (39.5 MHz) and FM sound (33.5 MHz) signals are present, and it is arranged that intermodulation takes place between these two. As a result, a *difference* frequency of 6 MHz is produced, containing modulation components of both contributing signals. Its AM (vision) modulation is minimised by a 'swamp' technique of depressing the 33.5 MHz sound carrier (in the early vision IF stages) to a level below that of the lowest (i.e. peak white) excursion of the vision signal. A tuned acceptor circuit picks off the post-detector

6 MHz beat signal for processing in an IC containing a series of amplifiers and limiters which remove any residual vision components, leaving a pure FM signal at 6 MHz containing sound information. After detection – again a synchronous, *quadrature*, detector, is used – the audio signal is back in baseband form, without any vision buzz if the limiters are doing their work! Amplification and processing in volume and tone controls render the audio signal in a form suitable for driving a loudspeaker.

In the MAC and Nicam receivers to be described later in the book, the audio signals, which as we know are in stereo, are produced in a totally different way. Here we have a strobed data-slicer which produces numbers corresponding to the quantised audio signal levels; they become recognisable after their passage through a D/A converter.

Sync separator

The detected video signal also feeds the sync separator to produce the all-important timing data for scan synchronisation, described in Chapter 1. This is a relatively simple circuit, internal to an IC, which discriminates between vision and sync signals on an amplitude basis, a sort of electronic bacon slicer, as it were. Usually a noise-cancelling circuit is provided in the chip which disables the separator during interference bursts, preventing picture judder and tearing. The sync separator is in the same package as the line oscillator, and TV receiver chips incorporate sync separator, line and field oscillators and several associated functions in a single IC package.

Timebases

The vertical and horizontal scans are produced respectively by the field and line timebases subjecting the electron beams in the picture tube to deflection forces by electromagnetic means, as explained in Chapters 1 and 7. Considerable magnetic force is needed to provide full deflection of the beams vertically and horizontally, and this force is generated in the field of the scanning coils on the neck of the tube. Figure 8.2 illustrates a typical deflection yoke and associated line output transformer for use with a colour picture tube.

Line timebase

The line timebase (Figure 8.3) consists of a group of building blocks; those which work at low voltage and current levels are incorporated in an IC. The 'pacemaker' is a pulse-generating oscillator whose frequency and phase can be controlled by an externally-derived voltage. It runs at line-scan rate, which is 15 625 Hz for 625-line 50-field systems and 15 750 Hz for 525-line 60-field systems. For a locked and stable picture display it must be synchronised with the broadcast line sync pulses (Figure 1.8, Chapter 1) not only in frequency, but in phase too. This is achieved by a *flywheel* synchronisation loop, whose function is to 'average' the

Figure 8.2 *Accessories for use with a 90° PIL tube: the deflection yoke and diode-split line output transformer*

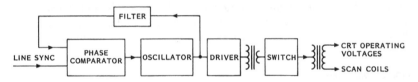

Figure 8.3 *Block diagram of the line timebase and scan-current generator of a TV set or monitor*

effect of the sync pulses over a relatively long period so that impulsive noise and interference in the TV signal cannot upset scan timing to give a ragged effect on vertical picture features.

Incoming pulses from the sync separator are checked against line oscillator pulses in a *phase comparator*, whose output voltage indicates by its polarity whether the line oscillator is running faster or slower than the broadcast sync pulses; and by its amplitude the difference in speed between the two. This error voltage is used to control the line oscillator frequency by 'steering' it to the point where there is zero frequency or phase difference between the two pulse chains, at which point the line oscillator is locked to the broadcast scan rate. The flywheel effect is provided by a low-pass filter – with a relatively long time constant – in the path of the DC control voltage feed to the line oscillator. In practice there are two loops, a coarse one to quickly bring the oscillator onto correct frequency, and a fine one to establish precision phasing of the line oscillator to the vision signal; adjustment of this phase loop, by manual or automatic means, determines the horizontal position of the picture on screen. For use with a tape or disc video source the time-constant of the flywheel loop filter is shortened, often by a sampling/detector circuit within the line oscillator IC; this prevents horizontal instability and twitch of the picture as a result of the timing jitter which is inevitably superimposed on the video signal by a mechanically-scanned readout system. In this circumstance the timing jitter affects both vision signal and sync pulse, so too long a time constant in the flywheel filter would smooth out the variations in scanning phase while the still-jittering picture information would give rise to a ragged and wavering image. A suitably short flywheel time constant permits the individual line scans to follow the random changes in pixel timing, thus straightening up the vertical features of the picture.

The line oscillator – whose basic timing element may be a quartz crystal, a ceramic resonator or a C/R combination – is designed to free-run in the absence of a synchronising signal so that the timebase continues to run (and generate EHT voltage) in the absence of an input signal or while tuning. The output from the oscillator section consists of a series of pulses with a repetition rate of 64 μs for a 625-line TV set or monitor. Throughout the rest of this chapter we'll assume that we're dealing with a 625-line 50-field system.

Having generated and synchronised a drive pulse for the line output stage, we need to provide impedance matching and drive power for the line output transistor, which in this role acts purely as a switch, as we shall see. It's usually provided in the form of a line driver stage, consisting of a medium-power transistor driving a step-down transformer as shown right of centre in Figure 8.3. On each line drive pulse the transistor saturates, inducing a pulse in the low-impedance secondary winding of the drive transformer. It is used to inject current into the base of the line output transistor for precise control of its conduction.

Line output stage
Although some practical line output circuits seem complex, their basic operation is very simple, and their essential elements are just four components as shown in Figure 8.4. Starting with zero energy in the coil and the picture-tube's scanning spot at screen centre, the switch S is closed, shorting out D and C, and applying the supply voltage across the inductor L. It's a basic characteristic of an inductor under these circumstances that the current in it builds up from zero in a linear fashion. After about 35 μs, and with the current in L still increasing, the switch S is

Figure 8.4 *Operating principle of a line output stage. In practice S is a transistor*

suddenly opened, whereupon the magnetic energy now stored in the ferrite coil of L is released, reversing the current flow and charging tuning capacitor C via large capacitor Cres, which is part of the power supply section, and may be regarded here as a short-circuit to the fast-moving currents involved in a line-scanning circuit. At the point where S was opened, the flyback of the scanning spots was initiated, and now they are back at screen centre and halfway through their rapid 'flyback' phase, during which the picture is blanked. The energy in L and its core is once more at zero, all having been transferred to C.

Continuing the action of an LC tuned circuit, the charge in C is fed back into L, and this exchange of energy between L and C would continue until it is all dissipated by the circuit losses were it not for the presence of clamp diode D, which comes into conduction after one half-cyle of LC oscillation, at the point where the bottom end of L attempts to move negative of ground potential. The magnetic energy stored in L now decays linearly to zero, feeding energy back into the supply line, and providing the first half of the next horizontal scanning stroke. When the scanning spots reach the centre of the screen once more there is zero energy in both L and C, at which point switch S closes again, initiated by the precision-timed line drive pulse, to start the complete cycle anew. It is repeated at intervals of $64\mu s$ to give continuous horizontal scanning of the electron beams across the face of the picture tube in synchronism with the broadcast picture. In practice, L is not the scanning coil; it is a board-mounted transformer serving many purposes. The scan-coils proper are connected in parallel with it via a DC isolating capacitor which also provides *S-correction* to compensate for the non-linearity of scanning speed inherent in the use of a flat-faced picture tube; and usually a *saturable* inductor (linearity coil) to correct for the resistive component of the copper-wound scan coils. Switch S is in practice an NPN or field-effect power transistor, sometimes incorporating diode D in its envelope. It has to withstand a very high positive collector voltage during the line flyback stroke, whose duration is about $12\mu s$, determined by the resonant frequency of LC (approx 42 kHz) and corresponding to the line blanking interval in the picture transmission shown in Figure 1.8.

The line output transformer (LOPT) is a useful source of energy for other parts of the TV set or monitor. Secondary windings feed the picture-tube heater, the first anode and, by means of voltage multipliers, the focus and final anodes, the latter at a typical DC potential of 25 kV. This is called EHT (extra high tension) and is produced by a *diode-split* extension to the LOPT wherein several separate EHT secondary windings are used, each forming a 'cell' in association with a rectifier diode and a reservoir formed by inter-winding layer capacitance. Each of these EHT cells, buried within the LOPT, typically generates 6kV DC and they are connected in series to yield an EHT potential appropriate to the tube size in use. Except for very small tubes this will be between 20 and 28kV. The focus voltage for the picture tube is also generated in this area by 'tapping off' the output of the first or second EHT cell for application to the *thick-film* focus and A1 control potentiometers, which are also built onto the LOPT. These supplies, along with other LT and HT lines derived from the scan transformer, have good *regulation*, benefiting from the close stabilisation which must necessarily be applied to the operation of the line output stage to ensure constant scan amplitude (picture width) in the face of changing EHT current demand with changes of picture brightness and mains supply voltage. Stabilisation is largely the function of the power supply section (PSU) to be described later in this chapter.

Thus the basic diagram of Figure 8.4 has grown up into that of Figure 8.5, which though still simplified is more representative of commercial practice. The component designations of the previous diagram have been retained to help in identification. For use with a wide-angle picture tube (110° deflection, generally in screen sizes above 53 cm/21") it is necessary to amplitude-modulate the line scanning current at field rate with a parabolic waveform. It's done by varying the resonant tuning of the scan-coil circuit – not the LOPT – at 50 Hz frequency by means of a *diode modulator* drive from a medium power transistor.

Field timebase

The function of the field timebase is simpler than that of its line-scanning counterpart. Its job is to deflect the scanning spots relatively slowly (cycle time 20 ms) from top to bottom of the viewing screen, then rush them quickly back to the top during the field blanking interval. During the downward stroke the line timebase draws over 300 scanning lines on the screen, and the timing is such that they are distributed evenly over the viewing area. The lines of one field are traced out in the spaces between those of the previous field to satisfy the interlace requirement explained in Chapter 1. The triggering of the field timebase by the broadcast sync pulses is critical for good interlacing performance.

Figure 8.6 is a representative block diagram of a field timebase system. It starts with a free-running oscillator capable of being triggered by the separated broadcast field sync pulses at 50 Hz rate. The timing pulses produced by the oscillator trigger a precision ramp generator whose output is a sawtooth waveform with a period of 20 ms and excellent linearity. It is amplified by a driver stage for application to the power output stage and thence to the field coils in the deflection yoke. AC and DC feedback are used respectively for linearity correction and stabilisation of the working conditions.

Field oscillators do not need a flywheel synchronisation system; they are triggered directly from the separated sync or make use of a digital countdown circuit, working from line rate, inside the timebase IC. As with the line oscillator, all that is required from the timing generator is a pulse, once per scan, to initiate flyback: the generation and shaping of the scanning waveform is carried out further downstream in the circuit, though not by the components associated with the scan coil itself at this low frequency at 50 Hz. As we have seen, the required scanning current is linear, with a rapid reverse of flow at the bottom of the screen, corresponding in time with the field blanking interval. The next stage in the field timebase circuit, then, is a sawtooth voltage generator. In practice it consists of a capacitor charged from a constant current source: this gives rise to a perfectly linear voltage ramp across the capacitor, terminated after about 19 ms by a shorting switch driven by the field oscillator. After about 1 ms, during which field flyback takes place, the short is removed and the ramp capacitor permitted to begin charging again. To avoid loading the capacitor its waveform is buffered by an amplifier with very high input impedance on its way to the driver and output stages, which are very similar to those of a medium-power audio amplifier.

Figure 8.5 *Practical line output circuit incorporating auxiliary voltage generators and horizontal pincushion distortion correction*

113

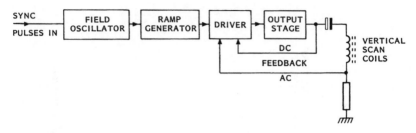

Figure 8.6 *Block diagram of field timebase section of a TV or monitor*

The power amplifier, now invariably incorporated into a heat-sinked IC, is made up of a Class B complementary-symmetry transistor pair whose mid-point feeds the deflection coils via a large DC-blocking electrolytic coupling capacitor. The DC feedback loop around the output stage stabilises operation in the face of temperature and supply-voltage variations, while AC feedback is used to set the gain of the amplifier and provide scan-linearity correction. It is taken from a low-value resistor in the ground-return path, which puts the deflection yoke itself within the loop and prevents picture height shrinkage as the coils warm up in use – the copper wire with which they are wound has a positive temperature coefficient.

Most modern practice uses separate ICs for generation/synchronisation of the sawtooth drive voltage and the driver/output stage. Figure 8.7 shows a skeleton circuit of an IC-based field output stage. The IC is powered from a 25 V supply

Figure 8.7 *Complete circuit of commercial field output stage using a single IC type TDA8170*

rail, derived from the LOPT, which enters at pin 6. During the scanning stroke C1 is allowed to charge to 25 V via D1, but for the duration of field flyback its negative plate (IC pin 3) is internally switched to IC pin 2. The effect is to reverse-bias D1 and 'jack-up' the chip's supply voltage at pin 6 to about 50 V to rapidly reverse the current in the deflection coils for a fast flyback stroke. This removes the need to continuously operate the IC from an (e.g.) 50 V supply and thus reduces internal heat dissipation and improves efficiency. Sawtooth current flows through the scan coils via coupling capacitor C2 and low-value (typically 1Ω) resistor R2 to ground. R2 acts as a sampler, developing a sawtooth voltage which corresponds to coil current. It is fed back to the driver stage via IC pin 1, having passed through an RC network incorporating height and linearity correction, often including S-correction to compensate for the necessity of scanning a flat-faced picture tube as described above for line scanning. In many sets provision is made for cutting off the tube's electron beams in the event of failure of the field deflection so that the screen phosphors do not get burned by the very bright horizontal line which would otherwise result.

Power supply section

A TV set or monitor requires a stabilised and closely regulated supply for its line scanning circuit to ensure that picture size and brightness remain constant as beam current, mains supply voltage and internal temperature vary. Other sections of the set also require closely-stabilised supply voltages, which may be derived from the LOPT itself or from separate feeds from the power stabiliser section: usually a mixture of both is used. The power supply unit (PSU) of a TV set, and indeed most other types of electronic equipment including VTRs, generally works from the domestic mains supply and provides one or more isolated supply lines whose voltage is held steady in spite of wide variations in the current drawn from them and the mains input voltage. Also required are protection circuits to prevent damage in the events of excessive current loading and excessive output voltage due to a malfunction in the stabiliser circuit.

Most PSUs use the 'chopper' technique, in which the regulating element is an electronic switch, as shown in Figure 8.8. The incoming AC mains voltage is rectified and smoothed in a mains-live area of the set to produce a DC operating supply of about 330 V for application via a high-speed switch transistor to the primary winding of a ferrite-cored transformer. Because the switching transistor is fully off or fully on at any one time, it dissipates very little energy and so runs coolly and efficiently. The ratio of on-time to off-time of the switch (*duty-cycle*) is varied to suit the load demand. When little energy is required the switch dwells in the on position for only a short time during each cycle of operation, briefly topping up an inductive and/or capacitive reservoir which supplies the load during the switch's off period. When the load's energy demand increases, for instance when the picture goes bright and pulls more beam current, a feedback/monitoring circuit automatically increases the duty-cycle of the switch so that it remains on for a longer period in each operating cycle and increases the energy fed to the reservoir. In this way the energy drawn from the primary power source (usually

Figure 8.8 *Operating principle of a switch-mode power regulator*

AC mains) almost exactly matches that demanded by the set from instant to instant, with minimal wastage within the PSU itself. The energy reservoir, whether in inductive or capacitive form, need not be large or bulky: the chopper-switch operates at frequencies in the region of 20 kHz so that small and light inductors and capacitors can be used.

In Figure 8.9 the incoming mains voltage is full-wave rectified by bridge rectifier D1–D4 to produce an unregulated supply of about 330 V across reservoir capacitor C1. This is switched to the primary winding of transformer T1 by electronic switch Tr1. Each time the switch closes a current ramp builds up in the primary winding of T1 and magnetic energy is stored in its ferrite core. When the switch is opened, say 50 µs later, the magnetic field collapses, transferring energy to the secondary side of the circuit: rectifier diodes D5 and D6 conduct the current flowing in the transformer's secondary windings to charge C2 and C3. The phasing of the windings is arranged to prevent conduction on the secondary side of the transformer when Tr1 is conducting. The loads are continuously supplied by reservoir capacitors C2 and C3.

All the secondary-side supplies are activated when the transformer's magnetic field collapses, so it's only necessary to monitor one of them for control and regulation purposes – this is normally the one that supplies the major proportion of the load. The output voltage here is monitored to provide feedback to the

Figure 8.9 *Skeleton circuit of IC-based switch-mode power supply with mains isolation in transformer T1*

chopper control circuit. A comparator checks the feedback voltage against a reference level, generally derived from a zener diode, to produce an error/control voltage proportional to the load. It controls a pulse-width modulator whose product is a pulse train of varying duty-cycle which governs the conduction period of the switch to suit the current demand. Thus is the stabilisation loop completed.

In most modern circuits the frequency as well as the duty-cycle is varied to suit different load demands. Many circuits are self-oscillating, with feedback from the transformer to the transistor to provide oscillation. Others (Figure 8.9) use a purpose-designed IC as the control element to provide current drive to the base of a bipolar transistor, or voltage drive to the gate of a power field-effect transistor (FET) which provides the switching action. The power transistor may be incorporated in the IC, resulting in a simple circuit with few peripheral components. The three resistors in Figure 8.9 merit a few words of explanation: R1 is a *surge-limiter* to restrain the inrush current when C1 charges at mains switch-on, protecting the rectifier diodes and mains-switch contacts. R2 initially powers the control IC, while pre-set potentiometer R3 is used to set up the stabilised output voltage: it would typically be used to adjust the supply voltage to the line output stage to a level of 115 V. Electrical isolation between live mains and the circuits of the TV is provided by T1 whose primary and secondary windings are separated by insulating layers capable of withstanding over 2 kV. This demands an isolating break in the feedback path, too, often provided by an *optocoupler* whose two highly-insulated halves are linked by an infra-red light beam, generated and monitored by a light-emitting diode (LED) and a photodiode respectively.

TV control systems

All but the simplest of modern TV sets have remote control and an internal system-control section, usually based on a microprocessor. It governs the tuning; programme selection; 'analogue functions' like volume, brightness and colour; standby switching; Teletext functions; and internal signal routing. A non-volatile memory IC (EEPROM) holds user-programmed data like tuning points, preferred analogue settings and, in some designs, factory- or installation-settings such as picture geometry, line phase and decoder/RGB operating points.

An overall view of a microprocessor-based TV control system is given in Figure 8.10. The microprocessor IC is an eight-bit type, custom-designed and mask-programmed for the application and for the specific TV model to which it's fitted. In common with all micros, its heart is a central processor unit (CPU) which consists of gates, inverters, adders and registers to store the results of calculations. Its operating program/instructions are held in a read-only memory (ROM) within the IC, while some random access memory (RAM) is provided for temporary storage of data and instructions. Data enters and leaves the IC by way of *ports*, an important one of which is the interrupt (INT) port, whereby the remote control instructions are passed to an internal serial-to-parallel decoder for instruction processing. Tuning is controlled by pulse-width-modulation (PWM) square wave output for integration into a DC control voltage: it's passed to the tuning control pin of the RF tuner.

Figure 8.10 *Microprocessor control of TV set functions. Communication is via the two-wire serial data bus shown on the right*

Most of the commands to and from the microprocessor pass via a two-way *serial data bus,* in which one conductor carries clock timing/synchronisation pulses, and the other a string of eight-bit serial data words consisting in order of address and command bytes. The address bytes indicate which of the several ICs on the bus is being called up, and then which of its internal functions requires to be accessed. The following data/instruction byte is fed into the appropriate register within the 'slave' IC, then if appropriate converted into an analogue level to conform to the command. At the end of each byte the slave IC acknowledges its receipt to the master, which is usually the control microprocessor. Because the bus is bi-directional and accessible to all the devices connected to it, provision is made for any of them to assume the role of master and call upon any other (including the microprocessor itself) to be slave – each IC has its own unique address in the form of an eight-bit identification code, the first part of the address sequence.

The same two-wire data bus is used for communication between the microprocessor and the memory IC in both reading and writing modes. The data (which for programme tuning consists of 13 consecutive bits) is passed after the memory chip and location addresses have been specified, confirmed and acknowledged between processor and memory ICs.

Remote control

The cordless control handset contains a pulse generator/encoder IC which generates a string of serial pulse data; each command on the handset has its own unique code, recognisable by the control IC at the receiving end. The pulse train is fed, in the handset, to a small power amplifier feeding one or more LEDs whose radiation is in the (invisible) infra-red spectrum. The chopped infra-red light pulses are intercepted by a photodiode on the TV's front fascia panel for reconversion to electrical pulses and amplification in a very high gain AGC-controlled amplifier before clipping and passage to the INT port of the microprocessor.

To round up this thumbnail sketch of the TV control system, Let's see what happens when two consecutive commands are sent by the viewer via the infra-red remote control. First, say, BBC2 is keyed. As the key is stroked the remote control IC comes to life, its clock oscillator starts up, and it strobes and samples the key matrix lines to ascertain which has been closed. Now it generates the appropriate pulse train and switches on and off the LED to send 'chopped' light pulses at about 38 kHz rate. At the receiver they are picked up and amplified in the infra-red pre-amp, squared and passed into the microprocessor IC of our Figure 8.10. The arrival of the first pulse interrupts whatever the micro is doing and diverts the data into a decoder where the command is interpreted. In this case the instruction is to access the section of the memory in which the tuning data for channel 2 is stored. A memory address is generated and sent along the serial bus to the memory (EEPROM) IC which passes the tuning data back over the bus to the micro-processor. On its receipt the control IC changes the duty-cycle of its tuning-drive pulse train to generate the precise DC voltage required to line the tuner up with the local BBC2 transmitter as pre-programmed by the installer or viewer. It then changes the (e.g.) on-screen channel indicator to CH2 or 02, then settles down to monitor the local keyboard and infra-red sensor for any further commands.

Let's imagine that the next command is to decrease picture brightness. The command is generated and sent by the infra-red hand-unit as before, though with a different code of course. On its receipt by the microprocessor it is decoded, then loaded into a register for temporary storage while the address of the luminance processor IC is generated, sent down the data bus and acknowledged. Now the address of the brightness data register within the IC is generated, sent and acknowledged. Finally, the data itself is sent in the form of an eight-bit (only six of them are used for this particular function) byte and loaded into the register. They are translated by a D/A converter to a DC control voltage – lower than before, since we are *decreasing* the brightness – which governs the operating point of the black-level clamp in the luminance channel. The picture brightness decreases accordingly.

Many TV sets are fabricated on a single printed-circuit board. An internal view of a typical TV set appears in Figure 8.11.

Figure 8.11 *All the electronics of this TV set are mounted on a single board. The power-supply section is at top right, with the line output transformer below it. At bottom left can be seen the tuner and receiver sections and the 21-pin AV (Scart) socket for interchange of signals and operating commands*

9 *Teletext and enhanced text*

Teletext is the oldest form of digital broadcasting technology, having been used in the UK for over twenty years, during which the decoder (an 'optional extra' in a TV set) has steadily shrunk in size and cost, though not in the complexity of the internal workings of its ICs or 'chips'. Teletext uses 'spare' lines in the broadcast field blanking interval to carry a digital pulse stream which when decoded can type out data on the TV screen in the forms of text and simple graphics like maps and diagrams. TV lines 7 to 18 in even fields and 320 to 331 in odd fields are used to carry the data, which can be seen as lines of twinkling dots above the 'live' picture if its height is artificially reduced.

Each text line carries a colour burst to maintain the colour decoder's subcarrier oscillator synchronisation, along with the usual line sync pulse and blanking interval. The line time given to text pulses is the same for that of picture signal: $52\ \mu s$. To prevent intercarrier buzz with FM sound systems the amplitude of the text pulses is limited to 66% of the peak white signal. In this two-state AM data transmission system, binary zero is signalled by the text signal hitting black level, and binary one by the signal rising to the 66% level, which represents 0.462 V with a standard 1 V peak-to-peak video signal. The pulses are not square: they are filtered before transmission and further rounded by the receiver's IF and demodulator circuits, and thus reduced to virtually sinusoidal shape by the time they reach the text decoder. The pulses are broadcast in NRZ (non return to zero) form, so that the signal remains high to indicate a string of binary ones and low when a string of binary zeros are being conveyed.

Text data format

The text data rate is 6.9375 Mbits per second, corresponding to 444 times the line frequency, though the pulses are not necessarily locked to line scan rate. A datastream representing 01010101 etc. throughout the line will thus correspond to a quasi-sinewave at 3.46875 MHz. At this rate there is room for 360 data bits on each TV line used for text conveyance: these 360 bits produce a single row of graphics or characters in the teletext screen display.

The 360 bits per line are grouped into 45 eight-bit bytes, the first five of which synchronise and prime the decoder, and provide an 'address' to indicate where the row is to be placed in the text display. The remaining 40 bytes define the characters in the row, which can thus have a maximum of 40 letters, numbers or graphic blocks. Blanks or gaps, e.g. between words, have their own code so that data is always present.

The text display has 23 rows, permitting a maximum of $23 \times 40 = 920$ characters to be displayed on a single page to form simple maps, diagrams and alphanumeric text. Upper- and lower-case letters can be produced, as well as figures, symbols and 'building blocks' for graphics, in any of eight colours formed by the three primaries, plus black and white.

The teletext code table is shown in Figure 9.1. Starting at the top left-hand corner, the first four bits define the row number and the next three the column number. The codes are thus used to determine the character so that capital T for example would be called up by byte 0010101 (bits 1 to 7), while byte 1111101 produces the symbol #. Note that to change from a capital to a lower-case letter only bit no. 6 changes, and that the codes for numbers are made by their binary equivalents followed by the code 011. The 32 codes in columns 0 and 1 are called control codes, and they determine whether the codes in the following columns represent alphanumeric characters or graphics and assign some 'attribute,' such as a colour, to the characters to be displayed. These control codes all have bits 6 and 7 at zero: they produce no display themselves, occupying a single blank space in the displayed row – usually in the background colour of the preceding character.

Protection

On its way from the transmitter to the decoder the text data is vulnerable to corruption by noise and interference, so a protection system is used, in the form of an odd-parity bit at the end of each of the seven-bit codes shown in Figure 9.1 so that each becomes an eight-bit byte. The parity bit added may be a binary one or a binary zero, such that there is always an odd number of one bits in the byte. This enables the decoder to check whether any bytes with an even number of ones have been received and reject them as being corrupt. This simple system has two limitations: it fails if an even number of bits in the byte are incorrect, and it cannot provide data *correction*. Even so, it is adequate for use with text symbols, where a minor error often goes unnoticed, and is corrected anyway when the page is cyclically updated.

Some of the text data requires greater protection than is afforded by this simple parity check. If the data conveying the page number is corrupt, the wrong page could appear in response to a viewer's request, while if the row address data is wrong the row will be printed out in the wrong place on the screen. To avoid this the bytes that carry this information and the real-time clock data are heavily protected by means of a *Hamming Code* in which every alternate bit is a parity bit: bits 2, 4, 6 and 8 convey data while bits 1, 3, 5 and 7 provide protection, as shown in Table 9.1. This permits single-bit errors to be detected and corrected by inversion. The penalty incurred is that the volume of information conveyed by the byte is reduced.

b4 b3 b2 b1	Col/Row	0	1	2	2a	3	3a	4	5	6	6a	7	7a
0 0 0 0	0	NUL[1]	DLE[1]			0		@	P			p	
0 0 0 1	1	Alpha^n Red	Graphics Red	!		1		A	Q	a		q	
0 0 1 0	2	Alpha^n Green	Graphics Green	"		2		B	R	b		r	
0 0 1 1	3	Alpha^n Yellow	Graphics Yellow	£		3		C	S	c		s	
0 1 0 0	4	Alpha^n Blue	Graphics Blue	$		4		D	T	d		t	
0 1 0 1	5	Alpha^n Magenta	Graphics Magenta	%		5		E	U	e		u	
0 1 1 0	6	Alpha^n Cyan	Graphics Cyan	&		6		F	V	f		v	
0 1 1 1	7	Alpha^{n}[2] White	Graphics White	'		7		G	W	g		w	
1 0 0 0	8	Flash	Conceal Display	(		8		H	X	h		x	
1 0 0 1	9	Steady[2]	Contiguous[2] Graphics	)		9		I	Y	i		y	
1 0 1 0	10	End Box[2]	Separated Graphics	*		:		J	Z	j		z	
1 0 1 1	11	Start Box	ESC[1]	+		;		K	←	k		$\frac{1}{4}$	
1 1 0 0	12	Normal[2] Height	Black[2] Background	,		<		L	$\frac{1}{2}$	l		‖	
1 1 0 1	13	Double Height	New Background	-		=		M	→	m		$\frac{3}{4}$	
1 1 1 0	14	SO[1]	Hold Graphics	.		>		N	↑	n		÷	
1 1 1 1	15	SI[1]	Release[2] Graphics	/		?		O	#	o			

Figure 9.1 Code table for broadcast Teletext. This is held in a ROM look-up table in the character generator IC of the text decoder

Row coding

As we have seen, the text page is organised in 23 rows. Of these the most important is row 0 (top row) because it contains the magazine and page numbers, the service name and the date and time – it is called the header row. To capture a page for display the viewer keys in the magazine and page numbers: when this keyed-in data corresponds with the transmitted code the required page is selected and displayed on the screen.

The data format for the header row is shown in Figure 9.2. First comes a clock run-in sequence consisting of two bytes carrying a series of ones and zeros, 101010 etc. It synchronises the decoder's bit-sampling clock oscillator. Next comes a framing-code byte whose purpose is to identify the start-point of the data bytes in the following pulse train. Its pattern is 11100100, chosen to give a reliable reset signal to the rest of the decoder and to be amenable to correction of single-bit errors by a simple check and correction circuit. The next two bytes, which are Hamming-protected, carry the magazine and row-address information.

Table 9.1 *Hamming coding table*

Message	Hamming coded bits							
	B1	B2	B3	B4	B5	B6	B7	B8
0	1	0	1	0	1	0	0	0
1	0	1	0	0	0	0	0	0
2	1	0	0	1	0	0	1	0
3	0	1	1	1	1	0	1	0
4	0	0	1	0	0	1	0	0
5	1	1	0	0	1	1	1	0
6	0	0	0	1	1	1	0	0
7	1	1	1	1	0	1	0	0
8	0	0	0	0	1	0	1	1
9	1	1	1	0	0	0	1	1
10	0	1	1	0	0	0	1	1
11	1	1	0	1	1	0	0	1
12	1	0	0	0	0	1	0	1
13	0	1	0	1	1	0	1	1
14	1	0	1	1	1	1	1	1
15	0	1	0	1	0	1	1	1

They are followed by page number and time-code bytes to enable the viewer to select a particular page at a specified time. Following these, the next two bytes provide information governing the whole page which is to follow: the control codes shown in columns 0 and 1 of Figure 9.1 for subtitle, update, news flash and similar functions, all with Hamming protection. Eight data bytes in the header row have thus been used, leaving 32 for the display of characters. Except for the page number, the header row text is the same for all the pages of any one transmission. It shows the name of the service (Ceefax, Skytext etc.) and the day, date and real time in hours, minutes and seconds.

The following rows have the same clock run-in and framing codes as the header row. These are followed by individual magazine and row-address bytes, which are the only ones to have Hamming protection in the 'ordinary' rows. Bytes 6 to 45 contain eight-bit codes for the text or graphics in the line. Each byte is transmitted with the LSB (least significant bit) first and the parity bit last.

Transmission sequence and access time

We've seen that each text line transmitted during the field blanking period can produce one row of text on the screen. 24 TV lines are required to produce the 24 rows of a single page of text data, then, and at the normal transmission rate of 12 text lines per TV field it takes two fields to transmit a single teletext page. At a field rate of 50 per second, 25 pages a second, or 1500 pages a minute can be transmitted. If 750 pages are being transmitted the worst-case access time will

CLOCK RUN-IN	CLOCK RUN-IN	FRAMING CODE	MAGAZINE AND ROW-ADDRESS GROUP	PAGE NUMBER (UNITS)	PAGE NUMBER (TENS)	PAGE 'MINUTES' (UNITS)	PAGE 'MINUTES' (TENS)	PAGE 'HOURS' (UNITS)	PAGE 'HOURS' (TENS)	CONTROL GROUP A	CONTROL GROUP B	CHARACTER BYTES
10101010	10101010	11100100										

Figure 9.2 *Data format for Teletext header row. Each block represents an eight-bit byte*

125

be 30 seconds and the average wait 15 seconds, assuming that the pages are transmitted repeatedly in sequence.

To maximise the user-friendliness of the system, high-priority pages such as indexes are transmitted more often than others. Also contributing to speed of access is a *row-adoptive* feature, in which the totally blank rows which occur in many text pages are omitted from the datastream. In addition to this, the text rows of several pages are interleaved, timewise, in the transmission to give a perceived reduction of the access time: this accounts for each row carrying its own magazine number. Subtitles, assigned to page 888, have the highest priority: they are slotted into the first available TV line after the cueing point. These techniques are limited in their effectiveness in that most of them shorten the waiting time for some pages while lengthening that for others. For the user the most effective reduction in page access time comes from the use of the *Fastext* system.

Text data memory and Fastext

There is no need to store the parity bits, so the RAM (random access memory) capacity required for a single page of text is 7 (bits) × 40 (characters) × 24 (rows), calling for 6720 bits, the minimum required in the decoder. Most decoders have a two-page storage capacity, in practice 2k × 7 bits, so that they can automatically capture the page following the one being displayed. If, for instance, page 183 is called up, page 184 is also captured and written into memory for immediate display. The action of the viewer stepping forward to 184 prompts the decoder to seek, acquire and store the data for the next page (185) in the now unused half of the memory, overwriting stored page 183 with the page 185 data.

Fastext decoders incorporate larger memories which can store the requested page and the next seven, automatically loaded in sequence by the decoder; as each page is discarded the next page in the sequence is written into the memory. This is based on the premise that viewers usually step through pages sequentially, and gives instantaneous new-page readout when they do. The broadcaster's text editor can enhance the Fastext function by anticipating the user's requirements and sending additional instructions to the decoder, based on the current page's contents, for extra page acquisition. It works thus: the bottom row of the text page has four colour-coded 'prompts' which may typically be news, sports, travel and weather. The remote control handset has four keys with corresponding colours. Pressing any one of them selects the magazine and page numbers simultaneously, loading the required pages into the memory for instant selection. If, for instance, travel is selected the on-screen coloured prompts change to air, sea, road and 'continue', the latter providing an escape from the travel menu.

A 'packet' system, in which additional data rows are transmitted but not necessarily displayed on screen, is used for Fastext data acquisition. The packets contain additional control data rather than characters and are sent in advance of the teletext pages (which now become packets 0 to 23) so that the control data arrives ahead of the page to which it relates. Packet 24 *is* displayed on screen as a row because it contains the Fastext prompts, while packet 27 provides the page-linking data. There are eight packets in the text specification, used for various purposes including linked character sets (primarily for other languages) and general-purpose data transmission. Packet 30 is assigned to PDC (programme delivery control) for videocassette recorders; the data it contains is captured and stored in a suitably equipped VCR, and used to initiate a recording at any required time on any channel which radiates PDC data. Its advantage is that it compensates for any unexpected alterations to the broadcast timing schedules.

Decoder operation

Figure 9.3 shows a text decoder in the simplest possible block diagram form. The first block gates the data pulses out of the video signal and decodes them. The next block selects the data requested by the viewer, according to instructions sent via the control logic block, and passes it to the page-memory RAM, whose write-in control signal also comes from the control block. Read-out from the RAM is again governed by the control block: it takes place at a slower rate, a complete field scan period being required for a page in the memory to be read out. It was written in during short 52 μs bursts at varying intervals.

The data from the RAM is fed to a character-generator ROM (read only memory) which uses it to produce the character graphics and symbols shown in columns 2 to 7a of Figure 9.1, and to give them the colours and attributes assigned by the control data stored with the character codes. The outputs from the ROM consist of RGB signals in serial data form, plus a blanking pulse train. They are fed to an RGB signal processing IC like that shown in Figure 6.6, where they would enter on pins 12, 14, 16 and 9 respectively for onward passage to the picture tube.

Data from the viewer's infra-red control handset is fed into the control logic section of Figure 9.3 in serial form, most often along the two-wire control bus described in the previous chapter. After serial-to-parallel conversion it's stacked in registers ready to act on the transmitted text data and the character-generator ROM.

Figure 9.3 *The basic elements of a text decoding system*

127

Practical decoder

A block diagram of a single-IC text decoder chip developed by Philips Components is shown in Figure 9.4. The analogue video signal from the TV's video demodulator enters at bottom left where its black level is clamped and the sync pulses stripped out to synchronise the clock which drives the internal timing chain. Now the video signal is converted into digital form (more details in Chapter 11) and passed to a data slicer, whose function is to produce chains of clean sharp pulses for the acquisition and decoding block, in which the required page is selected. Page selection data is passed into this block from the I2C interface, operating from the two-wire control bus described in Chapter 8. This IC has a single-page memory of 1.1k × 7 bit; other types have two-page memories or an interface block for connection to an outboard memory.

The single-page data is passed in parallel form to the character generator ROM which contains a look-up table to convert the 7-bit control data into serial pulses, locked to scan rate, which make up the numerals, characters and symbols on screen. They leave the IC at top right along with three other control signals for the RGB processing IC and one for the field timebase section: the latter de-interlaces the vertical scanning on pure text pages to eliminate interline flicker.

Memory extension

We've seen that there is little scope for increasing the number of teletext lines transmitted during the blanking interval, and that the Fastext system is only capable of reducing access time when pages are called up in a logical sequence. For fast access to a large and random selection of text pages it's necessary to provide a large data memory capable of storing scores or hundreds of sets of page data so that page selection, generally taking less than a fifth of a second, becomes a process of memory search and read rather than a wait for coincidence between requested and transmitted page numbers.

These large data memories are called Background Memory systems, and consist of an interface/data router and up to 4 Mbit of DRAM (dynamic RAM), capable of storing a maximum of 512 pages. The same decoding and display techniques as outlined in Figures 9.3 and 9.4 are used, and the data is systematically stacked in the memories as it is received. When a page request is made by the viewer, the DRAM is rapidly scanned in the full-field mode: time is saved by restricting the search to the memory area assigned to the relevant magazine and page header information. To ensure that the most up-to-date page is presented the memory is scanned in reverse order, that is from the most recently stored page to the oldest one. When the required page is located the memory is searched forward to read the data out to the one-page display memory within the text decoder IC.

Multilingual teletext

The advent of satellite transmissions which cover many countries simultaneously, and the acceptance of the UK-designed teletext system as a world standard (now

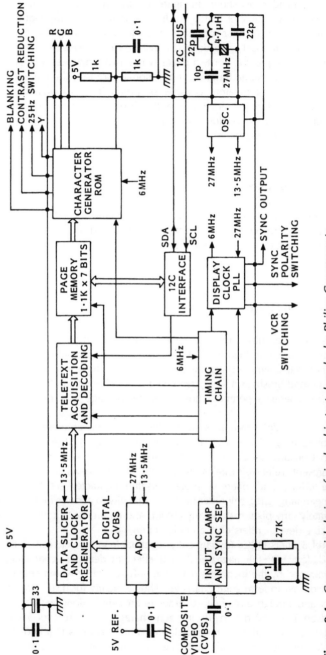

Figure 9.4 *Commercial design of single-chip text decoder by Philips Components*

129

used in over 35 countries) has led to the development of character generators capable of holding up to 192 characters, the extras being selected by the control bits in the header row shown in Figure 9.2 and by packet 26 which specifies the supplementary character that overrides the basic one at any required point in a row and column. Similar means are used to subtitle programmes in many languages simultaneously, the desired translation being selected by the viewer for insertion in the picture. For text pages where two languages may be required on the screen at the same time, e.g. educational purposes, there are ICs with dual-ROM character generators capable of producing two alphabets. Switching between them during a single row of the display is achieved by using a control character to shift the readout as required, in similar manner to the selection of alphanumeric or graphic characters.

On-screen display

With suitable software in the TV's control system the teletext ROM can be used to generate on-screen menus and captions for such purposes as programme/channel number readout, analogue control setting displays and viewer-programmed status titles.

Enhanced teletext

The system described above is called *basic alpha-mosaic*, and a typical page display is shown in Figure 9.5. While this is very acceptable for transmission of weather maps, simple diagrams and charts, they have necessarily to be drawn with a broad brush, as it were; certainly the possible total of 920 alpha-numerical characters per page permits a most adequate 'message' to be presented on a single page.

The basic alpha-mosaic system which we have at present can be greatly improved by getting more data into the page display. The CCIR (International Radio Consultative Committee) has proposed five levels of teletext, of which the present Teletext/Prestel system is the lowest. Level 2 is called *enhanced alpha-mosaic* and offers a more versatile character set to embrace most European languages, smoother mosaics, a facility to address individual text-pixels separately (any pixel may be any colour) and an increase in the range of colours available (level 1 offers only a choice of eight).

Level 3 offers DRCS (dynamically redefinable character sets) which are transmitted along with the magazine data. In effect the decoder's character generator is programmed by the incoming data, not only regarding which character to print, but also how to form each character. Thus a facility for displaying special characters and foreign alphabets to order is provided, along with better graphics, based on 12 × 10 dot matrix in each character cell. At level 3 these facilities can be used for either high-resolution or multi-colour as required, and it is possible for instance to provide quite high-resolution colour graphics over a limited area of the display.

Figure 9.5 *Off-screen photograph of a typical level 1 Teletext page broadcast by the BBC*

With such systems as the Canadian *Telidon* (text level 4) we move out of the area of alpha-mosaic displays and into another mode, *alpha geometric* where typically 75 000 pixels are available in the display, each of which can be separately addressed and assigned any one of the available 16 colours. For any given picture these 16 colours can be chosen from a quarter of a million available! Display of level 4 text calls for a degree of computing power and a great deal of storage capacity at the receiver if the full potential of the system is to be realised. As will be appreciated the method of encoding level 4 text is quite different to that for the alpha-mosaic systems, and each pixel is here defined on X and Y coordinates; the image is built up on the screen according to transmitted 'geometric' commands called *picture description instructions* such as point, line, arc, rectangle and polygon. An idea of the potential capabilities of level 4 can be gained from Figure 9.6 where the banana display contains 2600 bytes of data.

Finally, we come to text level 5. This is known as *alpha-photographic* mode and can, in its most developed form, offer a still-frame display better than that of any analogue-derived picture with a similar pixel count, which is limited only by the resolution of the display device, and the bandwidth of the transmission channel it occupies. As in level 4, each picture element is separately addressed, but now with 256 possible levels for each of R, G and B per pixel. This calls for the transmission and storage of a great deal of information, and would for instance require about $2\frac{1}{2}$ minutes to build up if sent over the currently-available text channel in a normal broadcast. To store the information for display, a page store at the receiver would require a capacity of 1.2 Mbytes (a byte is an eight-

Figure 9.6 *Level 4 graphics. Reproduced from an off-screen photograph supplied by the IBA*

Figure 9.7 *Level 5 enhanced Teletext. A full-definition frame with the capability of reproducing high-quality full-colour stills (photo courtesy BBC)*

bit *word*) which amounts to a full field store. Fill this store at 40 millisecond intervals and read-out at a suitable rate and we are back to moving pictures, digitally transmitted! Figure 9.7, which is reproduced from a BBC colour photo, indicates the results obtained in a level 5 extended teletext experiment.

This hierarchy of text levels can be engineered for compatibility between generations of decoders, so that the display is dependent on the sophistication of the decoder fitted to the viewer's set.

10 *PAL-Plus, MAC and enhanced TV*

With the coming of a new broadcast media such as cable and particularly satellite, whose bandwidth is less constricted than terrestrial broadcasts in the VHF and UHF spectra, much development work has been taking place on new encoding systems to provide displays free of the impairments described at the end of Chapter 4. Some offer the incidental advantages of potentially better definition and elimination of flicker and *aliasing* effects in the luminance picture.

Extended PAL

A relatively simple one, proposed by the BBC was the *extended-PAL* system, in which the inability of a conventional TV receiver to fully display the fine luminance detail (in the presence of colour) is acknowledged by transmitting a conventional PAL signal, but with the bandwidth of the luminance component restricted to about 3.5 MHz. This gives rise to little deterioration of the picture on a conventional set using PAL techniques and a shadowmask tube, where the presence of a luminance *notch-filter* and the shadow-mask (which acts to filter out fine detail, especially in the smaller screen sizes) virtually delete the finer detail anyway. Chrominance is transmitted in the conventional way, based on the subcarrier of 4.43 MHz, so that this scheme is a compatible one (see Figure 10.1), but without significant spectrum-sharing by the Y and C signals.

The high frequencies of the luminance signal, corresponding to fine detail, are sent separately, spaced from their rightful position by exactly the colour subcarrier frequency. Thus a 5 MHz luminance component appears at $5 + 4.43 = 9.43$ MHz. In suitably-equipped receivers the regenerated subcarrier can be used to drive a second luminance detector whose output, after filtering, can be added to the basic 'restricted luminance' signal to reinsert fine detail and achieve full definition pictures in colour without spurious effects. The gap in the extended-PAL spectrum between about 6 and 8 MHz is free for up to six digitally-modulated sound signals riding in a subcarrier (Figure 10.1 again).

This solution offers the advantages of compatibility and a low-cost receiving and decoding system for the enhanced colour pictures. It is, however, totally tied

Figure 10.1 *The spectrum of an extended-PAL system. High-frequency luminance components (section Y2) have been shifted up in frequency to largely eliminate Y–C band-sharing*

to the PAL system, and would have committed us to PAL and chroma subcarrier techniques for the foreseeable future.

Except in the sound channel, extended-PAL and similar systems take little account of the great advances which have been made in television technology in the decades since the subcarrier system was introduced into colour broadcasting, particularly in the area of colour picture encoding. With the economies of scale afforded by mass-production, digital and 'hybrid' ICs for use in television signal-processing can be made viable for domestic use in decoding systems which are more complex, but potentially capable of better performance.

PAL-Plus

A more sophisticated development of the PAL system, still maintaining the vital compatibility with existing equipment, is the PAL-Plus system. It offers the advantages of widescreen (16:9 aspect ratio) pictures, high-quality digital sound, and freedom from the cross-colour and cross-luminance effects (of band-sharing of Y and C components) inherent in conventional reception of subcarrier-based transmissions.

PAL-Plus pictures are seen in 'letterbox' format, with black margins above and below the picture, on ordinary receivers, because the analogue picture transmission contains 432 active lines. The others carry a digital 'helper' signal, used only in PAL-Plus receivers, which is used to build up a widescreen picture with 576 active lines. The PAL-Plus decoder has eight main building blocks as shown in Figure 10.2.

Analogue composite video (CVBS) signal from the TV's vision demodulator enters a conventional PAL decoder, the same in principle as that described in Chapter 6. The derived Y, U and V signals are A/D converted (see Chapter 11) into digital form for processing, and passed into a RAM capable of holding one complete picture frame: this calls for 6 Mbit capacity, or 12 Mbit in 'de-luxe' receivers with best-possible suppression of cross-colour effects. The RAM is under the control of a PTC (programmable timer controller), itself governed and instructed by a control data stream coming from the helper-decoder, whose function is to

Figure 10.2 *Block diagram of the picture-processing sections of a PAL-PLUS TV receiver*

extract the control information contained in the analogue video signal's 'picture-dead' lines. The data emerging from the RAM consists of eight parallel streams of digital information, sourced sequentially from different addresses within the memory store at high speed. It enters a sequential video processor (SVP), a real-time processor which converts the 432 active lines to 576-line display data by reference to the control data furnished by the digital helper signal.

Data from the SVP is passed to the next block, consisting of a widescreen signalling decoder (WSD), an eight-bit processor which uses a control signal carried on line 23 of the vertical blanking interval. Amongst other functions it ascertains whether a 4:3 or 16:9 aspect ratio picture is being broadcast and initiates automatic scan- and signal-switching. The processor also contains a freeze-frame facility, invoked by the viewer via the remote-control handset, which cuts off the flow of data to the RAM, whose contents are continually re-read at field rate, another technique which will be described in Chapter 11. The final block in the PAL-Plus decoder is the display processor (DP), basically a D/A converter to change the eight-bit digital data back to RGB analogue form to suit (after amplification by wideband high-voltage amplifiers of the sort described at the end of Chapter 6) the picture tube. The tube itself is conventional in every way except for the shape of its viewing screen. A TV set fitted with a 16:9 tube is illustrated in Figure 10.3.

Time-division multiplex

Where several signals need to be carried over a single transmission system, various possibilities are open to us. We have already explored the interleaving mode, which is only viable where the separate signals for dispatch have a common 'base frequency'. An alternative system is frequency-division multiplex, where the various signal channels are carried on different basic or carrier frequencies. Probably the simplest everyday example of this is the conventional TV aerial downlead which contains four TV programmes and accompanying sound tracks, all carried on different frequencies (channels) and easily sorted out by tuned acceptor circuits in the TV's front end.

135

Figure 10.3 *A widescreen TV receiver designed and manufactured in Europe by Philips*

A third alternative is the concept of *time-division multiplex*. Here the various streams of information are sent sequentially, each in a well-defined 'time slot' in the transmission. If we are to have a continuous signal stream available at the receiving end, this implies some form of storage system there to sustain the information flow during the periods when other information is present in the channel. A further requirement is that a synchronising signal be present at regular intervals in the composite signal to time the 'strobe' style detection and separation process.

Time-compression of analogue signals

Suppose we wish to send a five-minute spoken message over the telephone, perhaps to a far-distant country for which telephone time-charges are high. We may choose to compress our message into a time slot of $2\frac{1}{2}$ minutes by means of recording it on tape and replaying into the 'phone at double speed. If our correspondent (suitably forewarned!) were to hook the telephone to a tape recorder running at twice normal speed he could capture the message for subsequent replay at normal speed. In this way the *real-time* transmission phase is reduced by half, though at the expense of bandwidth. Inevitably the twice-speed message contains twice the frequency spectrum of the uncompressed message, and the media (in our case the telephone link) must be able to carry this.

Time-division multiplex and time compression are two ingredients of the MAC (multiplexed analogue components) transmission system. They are also used in the *digital* TV systems currently appearing in home TV and video equipment.

Multiplexed Analogue Components (MAC)

If we take a sample of a video signal about 700 times in each line period, and provide the means of storing the samples, we can 'hold' one TV line at a time in a memory, which can be analogue (charge-coupled device, CCD) or digital in nature, and embodied in a purposed-designed IC. We have to write in the samples at television scanning rate, but we can read them out at any rate we choose; see Figure 10.4. If we read them out at a faster rate we shall effectively have compressed the signal in time at the expense of a faster signal-rate, rather like our telephone and tape-recorder trick described earlier. The active picture period of a 625-line transmission is 52 µs (as shown in Chapter 1) so we shall need this period to write one TV line of luminance into the memory. Let's suppose we read it out in 35 µs, representing a time-compression of about 30 per cent. This will increase the data-rate proportionally, but leave a gap in the time-domain (assuming the standard TV-line 'window' of 64 µs) of 29 µs, in which we can do some time-multiplex magic to accommodate chroma and sound signals. Now for the chroma. Here we have about 18 µs available per line for transmission, and again we write the TV chrominance signal (say V, representing R – Y) into a one-line store. The read-rate here is faster than for luminance to squeeze 52 µs worth of V information into a real-time slot of 18 µs, time-compressing the V-signal by about 65 per cent, again with a proportional increase in data-rate.

The now time-compressed luminance and chroma signals are assembled in the form shown in Figure 10.5 which clearly shows the time-division-multiplex composition of the MAC signal. As only one colour-difference signal can be accommodated per data-line, V and U signals are sent alternatively on a sequential basis, rather like the SECAM system of colour encoding, where V and U signals are dispatched in the same way, though using a totally different modulation system, and operating in real time alongside the Y signal. Both SECAM and MAC require a

Figure 10.4 *Reading in and out of a TV line store at different rates. Each memory cell forms a short term store for a single pixel, and for a single TV line around 600–800 such stores would be required*

Figure 10.5 *The signal format for MAC. Full details are given in the text*

one-line delay at the receiving end, not for cancellation-matrix purposes like a PAL decoder, but rather to yield a stored chroma line while its fellow is being received.

MAC sound and sync

Our 64 µs data line is now occupied by about 53 µs of picture signal information in 'compressed' form, first 18 µs of V or U, then 35 µs of Y. This leaves about 10 µs for sound and synchronisation signals. While this may not sound very much, the tremendously-high bit-rate inherent in the system affords almost 200 bits in this period, which, remember, need only sustain the sound for a 64 µs period. Thus we have room in the *line sync word* for an initial eight bits (to provide system synchronisation for sound and vision demodulation), then sound and data (teletext) signals, all in digital form. The MAC-C system has potential for eight simultaneous data channels which may be used to carry either sound tracks or teletext pages.

MAC decoding

It is not possible to go into the MAC decoding process in much detail in this book, but as with any form of decoding, the process must be the 'reciprocal' of that at the sending end. Thus we write the luminance and chrominance signals into one-line stores at the fast transmission rate and simultaneously read out of them at normal 625/50 television rate to render continuous luminance signal and line-sequential V and U signals. By means of a 64 µs delay-line store and an electronic double-pole switch a constant flow of colour-difference signals is secured, each of R − Y and B − Y being updated at 128 µs (two-line period) intervals.

MAC interfacing

At the output of the MAC decoder we have sound and vision signals in baseband form. The latter may be RGB or Y, V, U signals, and ideally they should be fed direct to the picture-tube drive circuitry of the TV set (or monitor) in use to realise the full advantage of this mode of encoding. The MAC decoder can form an add-on unit to the TV set, or be an integral part of the down-converter for satellite reception, or the terminal unit of a wideband cable system. Where the TV receiver has sockets only for UHF or baseband CVBS inputs it is necessary to recode the chroma in PAL (or SECAM where appropriate in Europe) to enable the set to display a picture, which will necessarily then suffer all the shortcomings of cross-colour etc. related above! For TV sets with just a UHF input a suitable RF modulator as used in VTRs, TV games etc., will be required as an output device in the conversion equipment, whereupon a spare channel in the receiver will be reserved for cable and satellite transmissions, and tuned to the modulator's output. More details will be given in Chapter 23.

Subcarrier versus MAC signals

The three established encoding systems, NTSC, PAL and SECAM, were designed with the characteristics and particularly the noise performance of terrestrial VHF and UHF AM broadcast systems in mind, and with the need for compatibility with monochrome equipments at the top of the priority list. Many of the constraints which existed then have disappeared, and the main stumbling block – that of channel bandwidth – has considerably eased as carrier frequencies (and now bit-reduction ratios, see next chapter) have gone up and up. With the tremendous coverage area of satellite transmissions, a different form of compatibility now has to be considered, that of the different encoding systems – and languages – of the countries within the catchment area of a single transmitter in space, or indeed the vast distance over which TV signals can be propagated along fibre-optic light-guides. The language barrier is catered for by the availability of multiple sound channels, and the idea of component coding, in which each colour-signal is preserved intact throughout the transmission chain, gives great flexibility in the types and geographical spread of receivers in use, once the terminal equipment is freely and cheaply available.

Another advantage of MAC is the 'programmability' of its signal and synchronising systems, allowing for the first time in broadcasting formats a degree of flexibility to accommodate future changes and improvements in television technology, without rendering obsolete a great percentage of the total receiving installation.

High-definition and high-fidelity TV

The path to high-definition TV consists of several discrete steps, starting with improvement and enhancement of the existing display on current standards, and ending with the realisation of a TV picture with a wide-screen format and resolution capability equal to – or better than! – the best 35 mm cine film reproduction, and sound to match. As we shall see, the technical problems associated with high-definition TV are huge, but not insurmountable. Where home and broadcast TV are concerned, the old bogey of compatibility rears its head again, and many of the proposals being made are based on the idea that any new system will be amenable to existing receivers; an incompatible format would call for dual-standard receivers and possibly programme duplication, which is wasteful of spectrum space, no matter which media (or band) is used for transmission.

The digital MAC and PAL-Plus systems just described are the first steps towards enhancement of the TV picture. Although based on the traditional 625/50 system, the elimination of the worst visible defects of subcarrier colour systems such as cross-colour offers a great subjective improvement in reproduction. By incorporating two digital field-stores in the receiver, further improvements are possible. By interpolating between corresponding lines in the two interlaced fields and drawing more lines in the display, it is possible to achieve a subjective definition equivalent to a 900-line picture. The presence of a frame-store also facilitates reading out picture information at a faster rate than the 50 fields/second used in normal transmission to eliminate the flicker effect which is noticeable in large areas of high brightness. More of this in Chapter 11.

More than 625 lines

We have seen many times in this book that more picture detail calls for more bandwidth, and this is true whether the signal is transmitted in analogue or digital mode. For a worthwhile gain in picture definition we have to go beyond 1000 lines, and systems have been suggested using between 1100 and 1500 lines. A doubling of the European 625 standard to 1249 lines has often been advocated in the interests of compatibility and Philips have done much research along these lines. To draw 1249 lines in the same field period of 20 ms the scanning spot has to move twice as fast as before which means that to draw the same features the video signal must also move twice as fast, implying a doubling of bandwidth. Having done this, however, we are left with a picture which is capable of resolving much detail provided such detail is stacked vertically to take advantage of the higher line standard.

There is little point in doubling the vertical definition without a proportional increase in horizontal definition to match. To achieve this we need to double the vision signal bandwidth again, and so we arrive at a figure of 20-25 MHz for the luminance channel alone, or for each channel if the system is an RGB one. Such bandwidths are not easily accommodated, except with bit-reduced digital broadcast techniques, an idea of which will be given in the next chapter.

Aspect ratio

The aspect ratio of current TV displays is 4:3, largely because the first picture tubes had circular screens and the TV standards were based on these. For greater realism, it is generally accepted that an aspect ratio approaching 2:1 is a reasonable compromise between the ideal (which presumably is a cylindrical or spherical screen encompassing the viewer) and the practicalities of generating the display. Certainly the wide screen goes better with stereo and surround sound, and Hi-Fi TV will incorporate both these features.

High-definition TV programme sources

While 'electronic' signal sources will provide data displays and animated cartoons (along the lines of the text systems described in the previous chapter), most people are more interested in 'real' pictures, and these must come from cameras and ciné film. Solid-state image sensors have now been developed to the point where they can be used for picture pick-up in a high-definition TV system with the advantage of excellent RGB registration. Early experiments and demonstrations of high-definition necessarily used very-high definition camera *tubes*, based on the vidicon and Pumbicon principle, but usually of larger diameter, perhaps 30 mm.

The hardware for televising cine film is called a *telecine* and most use not camera tubes but photomultipliers for image-sensing, in conjunction with an FSS (flying-spot-scan) tube to illuminate, 'swinging torchbeam' style, the film frames as they pass the gate. At the very high data rate involved in high-definition TV neither photomultipliers nor FSS tubes are really fast enough, and experiments have involved CCD line-sensors in conjunction with a prism for pick-up (Philips) and a laser beam deflected by a rotating polygonal mirror as light source (NHK, Japan). It is fortunate indeed that good 35 mm (and especially 70 mm) ciné film contains sufficient detail to do justice to these new television systems!

Ironically, once high-definition TV is fully established in the professional closed circuit (as opposed to the broadcast) field, it may well signal the end of ciné film for professional purposes. Provided the necessary parameters are met, electronic film-shooting has many advantages over a conventional film camera in terms of versatility in subsequent processing; creation of special effects; correction of lighting and other deficiencies in the original scene; lower production costs in scenery and backgrounds, which can be subsequently 'added' by electronic mixing; and the ability to instantly review a sequence on a VTR monitor before sets are dismantled and personnel dispersed.

High-definition displays

The shortcomings of the shadowmask tube have already been discussed in relation to conventional television, and we saw that the shadowmask/triad combination sets a strict limit on the resolution available from this type of display. A well-set-up and magnetically-focused monochrome tube can display 1200 lines without too much trouble, but to achieve this with a shadowmask system is very difficult. There is another factor, and that is screen area. Even if we could achieve an 1149-line picture on a colour tube of conventional size (50-70 cm), it would subjectively show little resolution improvement over the present format, unless the viewer were within a few centimetres of the screen. For home use, a display area about 1.4m × 0.7 m seems optimum for high-resolution TV, and conventionally-scanned tubes are not very amenable, in a physical sense, to that size or shape!

To display our high-definition pictures, then, with sufficient brightness and resolution it is currently necessary to use projection techniques. For very large-screen applications one of the light-valve systems described earlier can be used,

Figure 10.6 *A projector for hi-fi TV by Philips. The lens system for the R, G and B tubes is clearly seen in this photograph. Other features are described in the text*

and the Eidophor system is demonstrably capable of handling pictures with more than 1000 lines. For smaller screens in the 1 m^2 region, direct-view tubes are now challenging projection systems in which the requirements of registering all three images within half a pixel all over the screen make tremendous demands on the convergence system.

Figure 10.6 shows an experimental hi-fi TV projector by Philips. It can define one million pixels, and uses three high-resolution 13 cm tubes running at 50kV EHT. It is intended for use with a concave screen having a light-gain of ×5. Dynamic convergence is handled by a microprocessor-based control system feeding separate correction coils on each tube.

High-definition TV summary

As our brief account has shown, the generation, distribution and display of high-definition TV pushes every aspect of TV design way beyond the limits imposed by the present TV system, in which the capabilities of all the links in the chain are fairly evenly matched. To bring them all up to standard for higher-resolution television pictures will cost a great deal of money, and require a great deal more research. Although all the technology for advanced TV already exists, and very impressive results have been demonstrated, the implementation of *domestic* systems (especially broadcast-derived ones) awaits many developments, mostly in the region of the economics of manufacturing practical fine-detail display devices.

The bandwidth problem has been solved by the development of digital TV broadcasting techniques incorporating data compression as we shall see in the next chapter. They are very complex, but are nevertheless suitable for home TV use because of the economies afforded by mass-production of the VLSI (very large scale integration) ICs involved. All the time that the conventional thermionic picture tube provides the display, however, there will be a need for the timebases, power supplies and shadowmask systems described in previous chapters.

3D TV

Perhaps the ultimate step in television technology is the presentation of a three-dimensional display with depth and perspective as well as full colour and high definition. One day this may be realised by means of *holographic* techniques in which interference patterns from laser light sources are able to create an apparent three-dimensional image without the need for the viewer to wear special spectacles or other optical aids. 3D systems using more conventional display sources have been demonstrated, however, and have even found their way onto domestic TV screens in the form of demonstration transmissions, for which special bi-colour viewing glasses are required.

One system uses two conventional TV sets, fed with video signals from two cameras viewing the same scene from different angles. The sets are mounted in 'L' formation with the screen planes at right-angles; each screen projects an image onto a sheet of reflective plate glass set at an angle of 45° between the two screens. Viewed through special spectacles the effect is very striking, with the picture appearing to project forwards into the space in front of the display surface.

An alternative approach, based on the stereoscopic principles known since Victorian times and exploited in some cinema films from time to time, involves a colour-separation process in which red images are separated from green (or cyan) and presented side-by-side on the display screen to render a picture which looks like a badly-converged colour display. The horizontal separation of the two coloured images determines the apparent depth and spatial position of the object thus outlined, and will vary to describe 'Z axis' (i.e. on a line between the viewer and the display surface) movement. This effect can be exploited in a basically black-and-white or colour picture.

A means of artificially inserting a 3D effect into ordinary colour broadcast (or videotape) pictures was introduced some years ago by European setmakers Saba and Nordmende. When invoked, the 3D facility horizontally separates the red and green images by means of an electrical time delay, and a quasi-3D effect is seen when viewing the result through the red/green spectacles supplied. There are several alternative approaches to 3D TV short of the full holographic treatment, and if and when a common and compatible system emerges it may well involve the use of polarised light.

11 Introduction to digital TV

Long before digital vision became a reality, digital sound was well established. First introduced as a domestic technology in the form of Compact Disc in 1982 by Philips, it had been used by broadcasters for point-to-point transmission in the Sound-in-Syncs system for some years before that. Digital vision techniques have been harder to implement because of the sheer volume of fast-changing data involved, calling for wideband transmission channels, large memory stores and very large-scale integration ICs, all of which have only become available, in a form compatible with home TV systems, relatively recently.

The digital video signal

The concept of quantising analogue waveforms was briefly described in Chapter 1, and following on from that we must now look in more detail at the way in which a TV picture signal is encoded in binary form. Figure 11.1 represents the video waveform over a period of 1 μs which, on a typical domestic TV with a 56 cm diagonal screen, would occupy about 9 mm of *one* scanning line. The picture pattern on this section of the scanning line can be seen to start at mid-grey, brighten up, then pause at 'bright grey' before dropping to a lower level. At the end of our 1 μs period the brightness is increasing again towards peak white. On the left-hand scale of the diagram is shown the quantising levels between 0 and 256, while along the bottom scale time intervals are shown up to 1 μs. At regular intervals of about 74 ns (ns, nanosecond, one thousand-millionth of a second) the video waveform is sampled to establish its level at that instant. In Figure 11.1, then our samples at 74 ns intervals are 143, 160, 223, 241, 234, 200, 196, 203, 175, 105, 103, 140 and 170, making 13 samples. These numbers are produced in a fast A/D (analogue-to-digital) converter, not in the form of our decimal numbers, but in binary code; an eight-bit (bit, binary digit) word is used, with the necessary range of decimal 0–255. Binary coding is the simplest way of representing decimal numbers in 'pulse' form; reading from the *right* the 1s represent the numbers of 1s, 2s 4s, 8s etc. Our first three samples, then, are 10001111, 10100000 and 11011111, and so on throughout our 1 μs 'window'. The digits come out in serial form, necessarily fast! All eight digits must be produced in the 74 ns period, so that for real-time digitisation we must expect around 100 bits per microsecond, repre-

Figure 11.1 *Quantising a video signal: the waveform represents a small portion of the video signal of one TV line. Quantising levels are represented by the crosses*

senting a bit-rate of 100 Mbits/second, though the signal is now in binary form, and has only two states, 1 and 0. Various techniques can be used to reduce this bit-rate without impairment, as we shall see.

There are two approaches to quantising video signals. Either the complete video signal (encoded, with subcarrier, to PAL, NTSC or SECAM standard) can be quantised at a high sampling rate to render a composite digital signal, or (better) each of the Y, V and U signals can be separately quantised to give three digital pulse streams for combination before transmission. In this system, called *component coding,* the sampling rate for each of the colour-difference signals need only be about half that for the luminance signal due to their lesser bandwidth, and correspondingly lower *Nyquist* frequency. Because colour-difference signals, as we have seen, can have positive or negative values, zero is taken as level 128 from which the colour-difference signal can go down towards zero for negative signals and up towards 255 for positive signals. The total of 256 levels available imply an eight-bit word as before. The advantage of component coding is the compatibility of the system, in which each component, Y, V and U is effectively kept separate throughout the transmission and reception chain. Upon receipt, they can be fed separately to the RGB matrix and display channels of a colour receiver, or encoded in whatever form is required for the country or area intended.

Back, now, to our quantised video signal of Figure 11.1. At the end of the chain the digital pulse stream is presented to a D/A (digital-to-analogue) converter, whose output can describe 256 different levels, corresponding to the numbers conveyed in the bit-stream. This is the recreated facsimile of the original analogue signal, and at the sampling rate employed, it is a very good analogy indeed, as shown in Figure 11.2. Over our 9 mm section of a single scanning line it is subjectively indistinguishable from the original signal, except for the complete absence of noise and other spurious effects! It is not hard to see that noise (and virtually all the noise – or snow – we see on our TV screens is picked up along the

145

Figure 11.2 *The video signal of Figure 11.1 reproduced after passing through a digital coder. It is a very good facsimile of the original signal*

way) would not be reproduced on the digitally-transmitted picture; provided the 1s and 0s of the pulse stream were distinguishable from each other in the receiver's decoder, no noise at all would be visible in the displayed picture.

A/D conversion

The A/D converter used for this relatively high speed application is known as a 'flash' type, shown in Figure 11.3. It produces an eight-bit output, using 256 separate comparators. Each of them has two inputs, one of which is connected to its own tapping point in a chain of 256 equal-value resistors, while the others

Figure 11.3 *Fast-acting A/D convertor of the type used with TV picture signals*

146

are all connected together and to the video input signal. A stable reference voltage, typically 2 V, is maintained across the resistor chain so that each resistor drops one 256th of the total voltage. The video input is gain-controlled to exactly 2 V peak-to-peak while its black level or sync tip is clamped at a fixed level such that the peak white video level just turns on the topmost comparator in the chain. As the video signal swings from black through grey to white it turns on progressively more comparators in 256 steps of 7.8 mV (0.0078 V).

When a clock pulse arrives the outputs (0 or 1) of the 256 comparators are loaded into a 256-to-8 binary encoder which converts the quantised samples into eight-bit bytes for passage to a latched buffer. On each clock pulse a fresh sample is taken and a new eight-bit byte produced. The frequency of the sampling clock varies with the application and the required picture quality, and may range from 13.5 MHz for the sort of picture definition we are currently used to in broadcast reception, to 30 MHz or so for high-definition systems. To avoid 'aliasing' and consequent spurious effects, a Nyquist filter limits the incoming video frequency to less than half the sampling clock rate. At 13.5 MHz we get 676 samples per 52 µs active line period: increasing the clock frequency or omitting the line blanking interval permits a greater number of samples to be taken. In a component coding system the luminance (Y) signal would be processed in the way just described, while the U and V components are separately dealt with in their own A/D converters whose principle of operation is identical. Since the bandwidth of the colour-difference signals is only about a quarter that of the Y component they require less frequent sampling. In practice they are often sampled at the same rate as the luminance signal, after which three in four samples are discarded by clocking the data out of the shift register at a quarter of the basic rate, i.e. 3.375 MHz for a Y sampling clock frequency of 13.5 MHz.

Thus for every eight-bit luminance byte there are two bits of U data and two bits of V data, complete U and V bytes being produced in the time taken for four luminance bytes. If the bit streams are interleaved in time (time-division multiplex, see previous chapter) a complete YUV 'bundle' consists of twelve bits as shown in Figure 11.4. The bit interleaving is carried out by *formatting logic* between the A/D converters and a 12-bit register.

Now that we have got the vision signal into digital form, having only two levels, it is amenable to storage in a digital memory IC, and to transmission over links whose only requirement is the ability to pass a high-speed string of 0 and 1 pulses with sufficient discrimination between them to permit recognition of 'on' or 'off' at the receiving end. Thus high noise levels, for instance, can be easily tolerated,

D0	D1	D2	D3	D4	D5	D6	D7	D8	D9	D10	D11	BITS	
Y0	Y1	Y2	Y3	Y4	Y5	Y6	Y7	U0	U1	V0	V1	WORD 1	13·5MHz
Y0	Y1	Y2	Y3	Y4	Y5	Y6	Y7	U2	U3	V2	V3	WORD 2	3·375MHz
Y0	Y1	Y2	Y3	Y4	Y5	Y6	Y7	U4	U5	V4	V5	WORD 3	
Y0	Y1	Y2	Y3	Y4	Y5	Y6	Y7	U6	U7	V6	V7	WORD 4	

Figure 11.4 *Interleaved luminance and chrominance data make 12-bit words to fully describe the characteristics of each picture element*

with no picture disturbance at all. The signal now consists merely of a 'blueprint' or 'typing instructions' for the picture, which is built up anew in the receiver, specifically at the DA converter to be described later in this chapter. First we shall examine the storage of a digital signal in a large RAM, and take as our example the display of a flicker-free picture in a domestic TV receiver.

100 Hz pictures

We have seen how the PAL-Plus system uses digital technology to improve reception of analogue TV broadcasts. Two impairments of conventional PAL/625 pictures are interline flicker, where sharp horizontal edges of objects in the picture and fine horizontal lines have a 25 Hz repetition rate; and large-area flicker, which is particularly noticeable with large screens and bright pictures. It arises from the inability of the human eye and brain to fully integrate successive pictures with a 'flash rate' of 50 Hz as currently used in Europe. In the USA, Canada and Japan where a 60 Hz field rate is standard, the flicker effect is lessened; at a flash rate of about 90 Hz it disappears under all picture brightness, viewing angle and ambient light conditions. The use of a large RAM IC storage system removes the need for the picture to be scanned in real-time synchronism with the transmission and opens the way to the use of a 100 Hz display for flicker-free viewing.

The basic idea is shown in Figure 11.5, and is simple in concept. The video signal is converted from analogue to digital form and written into the memory at the 50 Hz broadcast rate. It's read out of memory at the faster rate of 100 Hz, converted from digital to analogue form and displayed on screen at double-speed line and field scan rates.

Memory ICs

The field store memory generally consists of three 1 Mbit fast DRAMs (dynamic random access memories) with sequential addressing, each arranged as 256K × 4 bits. They are connected in parallel, with one IC storing the four least significant bits, the second the four middle bits and the third the four most significant bits of the 12-bit time-division multiplex YUV data shown in Figure 11.4. There are several alternative ways of storing the data: eight-bit Y data can be stacked in its own 2 Mbit (e.g. 2 of 4 bits × 256K) section while the four-bit UV data is held separately in another 1 Mbit IC; or three 2 Mbit stores can be provided, one each for Y, U and V data. While this latter system requires an expensive 6 Mbit capacity it provides a YUV sampling ratio of 4:4:4 and full bandwidth in all three channels, which is useful for RGB operation.

With the 4:1:1 ratio scheme the RBG inputs have first to be filtered then matrixed to obtain YUV form and then stored for reconversion to RGB format after D/A conversion in the 100 Hz part of the circuit. This is not such a problem as it seems: no encoding, modulation or interleaving is required, and with an RGB signal most of the information in the three channels is common to them all, and thus redundant, as we saw in Chapter 3.

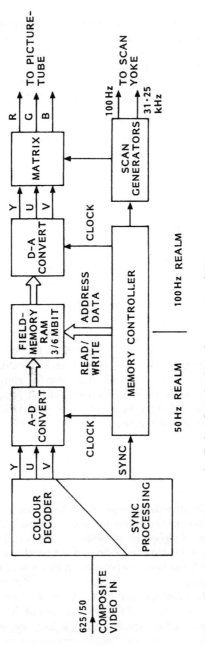

Figure 11.5 *100 Hz processor for flicker-free display of conventional TV pictures*

Digital noise reduction (DNR)

The presence of a picture field store in a TV receiver or monitor opens the way to a noise reduction system. There are many reasons why noise intrudes on an analogue video signal, amongst them weak signal reception and tape noise in VCR playback. Noise is random in nature, with no correlation between successive video fields, while those of a still video picture have total correlation. Thus if several video fields are integrated the noise cancels to virtually zero while the picture-signal components add.

Electronic noise reduction can be effected in a field-store TV set by progressive integration of the present and previous fields. Figure 11.6 shows the basic idea in block diagram form. The block marked K is a *recursive filter* whose characteristic

Figure 11.6 *Noise reduction by successive integration in the digital realm*

determines the noise-reduction factor: the higher the K factor the greater the noise reduction. In practice the recursive filter is governed by a motion detector which continually adjusts the K factor, reducing it for moving parts of the image (to prevent blurring due to integration) and increasing it for stationary picture areas. The effect of this is that the longer an object in the picture remains still, the better and more noise-free it appears.

Memory readout

A memory controller IC governs the read and write processes and generates the sync pulses for the 100 Hz/31.25 kHz timebases. Figure 11.7 shows in block diagram form a system designed for domestic TV sets which uses two sets of field memories, DNR, 4:4:4 sampling ratio and field and line flicker reduction. The sampling and read/write rates can be selected by the setmaker or in software to suit different scanning rates and standards. At the higher rates the picture definition is greater than that required by the CCIR standard, but all rates call for very short DRAM access times. If the time-constant of the voltage-controlled oscillator (VCO) controlling the memory readout is long, the system takes on the characteristic of a timebase corrector (TBC) ironing out timing jitter in the video signal. This will, for example, provide better off-tape picture stability. Plainly it's not possible to combine this with effective DNR because the pixels in successive fields are no longer time-coincident.

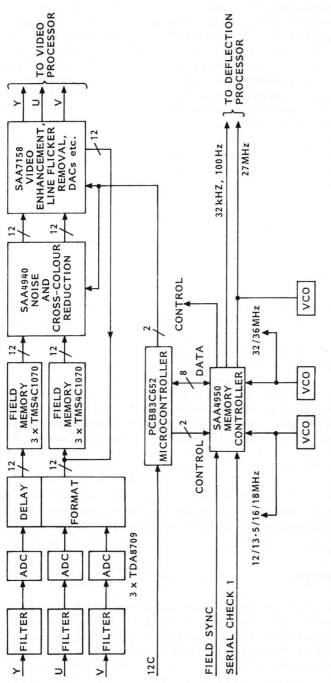

Figure 11.7 *Digital picture processing system for 100 Hz display of broadcast TV pictures. Based on a chipset by Philips Components*

151

Some other features of the system shown in Figure 11.7 will now be described. The delay in the Y data stream is part of the bit interleaving/data formatting system which produces the 12-bit YUV words of Figure 11.4. The SAA4940 IC carries out noise- and cross-colour reduction by recursive filtering. Downstream from it the SAA7158 device implements interline flicker removal, U and V signal reformatting, digital colour transient enhancement (which subjectively sharpens up vertical edges of coloured picture objects) and Y signal peaking. It also contains the necessary high-speed DA converters to bring the signal back to analogue form as shown on the right of Figure 11.5. Two additional features of this IC are called Zoom 1 and Zoom 2. The former provides vertical picture expansion so that 4:3 aspect ratio letterbox pictures can be correctly displayed on a 16:9 aspect ratio picture tube. Zoom 2 offers magnification by two of the displayed picture, both horizontally and vertically, as a viewer option: in effect each pixel is quadrupled in size in the 'magnified' picture. All the memory-control and post-memory processing functions can be software controlled via a serial bus which is also governed by the user's infra-red remote control handset.

Special effects

In addition to the noise-reduction and picture zoom features described above, various other possibilities exist in a system incorporating a digital field store. An obvious one is a perfect freeze-frame facility: the memory writing process is halted while the 100 Hz readout continues as long as the viewer requires. Other possibilities are *pixelation*, in which the D/A converter clock rate (see later) can be progressively slowed – in practice progressively halved – by the viewer to give the picture an increasingly mosaic-like appearance; and *solarisation*, where the two, three or four least significant bits are removed from each byte. This reduces the number of analogue signal levels, giving the reproduced picture a surreal effect.

Where the field-store memory bank is part of a video effects console rather than a TV set, further possibilities are exploited. It's possible to squeeze, zoom, reverse and 'wave' the picture by changing the memory reading rate and direction; to superimpose two pictures whose original scanning generators are not locked in sync with each other; to provide *chroma-key*, in which another picture is inserted over (e.g.) all blue areas of the main image; and so on, depending on how much computing power is available in the system- and memory-controllers.

D/A conversion

When all the tricks and manipulations have been carried out on the digital data containing the picture (and this may include compression, transmission and reception, see later) it's time to reconvert the data into analogue form to suit the requirements of the display device, be it a picture tube or a solid-state panel.

Where necessary the digital data stream is de-interleaved to separate it into its Y, U and V components for application to three separate D/A converters. Each of them works on the principle shown in Figure 11.8. The first step is serial-to-

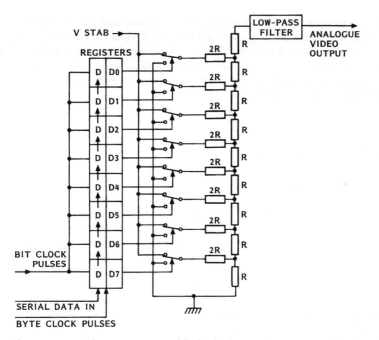

Figure 11.8 *D/A converter suitable for high-speed processing of video data and signals*

parallel conversion: the eight-bit serial data words are stepped through a register by the bit-clock pulses at the memory read rate, typically 27 MHz for 100 Hz TV display or 13.5 MHz for domestic 50 Hz applications. The eight-bit words are transferred to a slave register at the byte clock pulse rate, thus making parallel data words available to the D/A converter proper. Each bit in the word being converted controls a two-way switch connected to an R/2R ladder network with eight 'rungs'. The ladder has a closely-stabilised voltage applied across it. The effect of each bit in the word depends on the position of its switch down the resistive ladder. Thus the most significant bit (MSB) D0 at the top of the chain commands 50 per cent of the output voltage swing, the next significant bit D1 25 per cent and so on down to the least significant bit (LSB) D7 which influences the output voltage by a mere 1/256th or 0.4 per cent.

A 256-step video signal cannot, when viewed on screen, be distinguished from a non-processed analogue-derived one by the human eye. The other advantage of eight-bit processing (it's possible to get away with six- or seven-bit quantisation) is that the noise introduced by the digital processing is very low at −54 dB, virtually imperceptible in normal viewing. In the 100 Hz picture system described above a low noise figure is essential for a high quality picture.

D/A conversion is followed by a sharp cut-off analogue filter (usually an LC type) to smooth out the quantisation steps and suppress the remnants of the clock waveform. The Y filter cuts off sharply at a frequency rather less than half that of the memory-read clock, say 12 MHz for a 27 MHz readout rate, and 6 MHz with a 13.5 MHz rate. Since the frequency of the U and V signals is so much lower than the D/A clock rate their filter requirements are much less demanding. To capture all the available picture detail the bandwidth of the analogue circuits downstream of the Y converter, and in the RGB channels, must reach up to the transition frequency of the Y filter.

Digital picture broadcasting

The stumbling-block, for many years, in digital television transmission has been the very fast bit-rate involved in fully describing all the detail in every pixel at the rate of 50 fields per second. The bit-rate for pictures of the quality we are presently accustomed to is 216 Mbit/second, corresponding to a sampling rate of 13.5 MHz and a 12-bit word expressing the luminance and chrominance values in 8 + 2 + 2-bit form as described earlier in this chapter. High-definition TV gives rise to an even faster bit rate: for a 1250-line, 50-field, 16:9 aspect ratio picture data is generated at the rate of 1.2 Gbit (1200 Mbit)/second. While these rates can be handled in studio and closed-circuit applications (digital operation has been standard in some control and editing suites for many years) they are quite incompatible with current and future broadcast and cable distribution systems, whose bandwidth is limited and costly. To reduce the bandwidth and storage capacity required for video datastreams, two techniques are used: bit-rate reduction, and what might be called economical modulation systems. Each will be outlined in turn.

Bit-rate reduction

An alternative term is *digital compression*, whose object is to reduce the bit-count as far as practical while facilitating the re-creation of the best possible picture at the receiver. It is made possible by the fact that most television scenes contain much redundant information. This redundancy takes two forms: *statistical*, in which picture and pixel samples are closely correlated with adjacent samples, both in the same frame and in previous and successive frames; and *pyschovisual*, in which advantage is taken of the tolerance of the human eye and brain to certain types of distortion in the image, dependent on its context and content. Bit-rate reduction removes the statistical redundancy and distributes any remaining distortion in time and space on screen so that its visibility is minimised. Some scenes have great redundancy: a still image, or stationary part of a picture, needs sending only once and can then be repeatedly read out from a store at the receiving end. Imagine a scene consisting of a crowded beach beneath a blue sky. The sky section of the image has almost total redundancy, so that data capacity 'saved' in transmitting the sky image can be 'spent' in reproducing the detail and movement of the lower part of the picture. Plainly the most difficult images to reproduce by this compression system are ones that

Figure 11.9 *Principle of bit-rate reduction as used in digital TV systems*

are highly-detailed and rapidly changing: examples are jump-cuts between 'busy' pictures; and camera zooms amid sharply-defined images.

The bit-rate chosen for transmission depends on many factors. To convey a picture subjectively equal to that of a standard (low-band) VCR replay image, a rate of about 3 Mbit/second is adequate. At progressively higher rates the reproduced picture improves. Figure 11.9 outlines the basic bit-reduction technique. The incoming fixed-rate high-density data passes first through a stage which removes the statistical redundancy to render a datastream whose bit-rate varies according to the characteristic of the video signal: as we have seen, some pictures (and parts of them) have more redundancy than others. To match this varying data rate to the fixed rate transmission channel a buffer store is provided, typically with a digital storage capacity of two frames. The buffer store acts as a regulator with a constant rate of output data. Its content is continuously monitored, and when it nears overflow a 'slow-down' signal is fed back to what might be called a distortion processor whose purpose is to reduce the buffer's input rate and thus maintain equilibrium. The manner in which the distortion is introduced is very critical: it's based on an algorithm carried out in a discrete cosine transform (DCT) processor, a microprocessor-based block capable of performing foward and reverse transform operations in real time. It processes the data in 'chunks' corresponding to 8 × 8 pixel blocks to produce coefficients representing the amplitudes of the various frequencies present, and the DC level of the whole block.

By scanning the data matrix in a controlled manner, called the *matrix transform*, long runs of zero and near-zero values are generated, reducing the amount of significant data to be sent. The remaining coefficients are quantised to reduce the number of non-zero values in a technique called *adaptive quantisation*. The number of bits in each sample is further reduced by using a 'shorthand' coding system based on the work of Huffman, in which the most frequently occurring symbols are allocated the shortest possible codes. Finally *run length* codes are used, in which, for example, the sequence 3,3,3,3, can be coded 3,4 (3, four times) to minimise the number of symbols transmitted. By these means compression ratios of 50:1 and more can be achieved without significant subjective deterioration of the reprocessed picture. At the decoder the process is reversed, with the help of a two-frame digital picture memory store and suitable processing circuitry operating from a look-up table containing the 'codebook'. The digital data recreated in the decoder is D/A converted and reproduced on screen as a picture with excellent quality and virtually imperceptible noise.

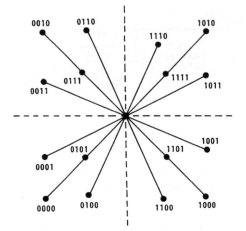

Figure 11.10 *16-QAM modulation system showing vectors and amplitudes for conveyance of four bits in each transmitted symbol*

Modulation of digital picture data

The datastreams representing control, data, sync, video and audio (see next chapter) are interleaved in time, and now have to be impressed onto a carrier for transmission. Several efficient modulation systems have been devised for different carrier frequencies, media and circumstances.

Figure 11.10 illustrates a method by which sixteen different four-bit binary codes can be represented by 12 phases of the (single-frequency) carrier wave; four of them, widely separated, have two possible amplitude levels making 16 states in all. It's called 16-QAM modulation, and as techniques improve, particularly in terms of accurate passage in the receiver and recognition at the decoder, it becomes possible to develop this technique to the point where eight-bit symbols are conveyed in 256-QAM modulation systems.

OFDM

Orthogonal frequency division multiplex (OFDM) modulation is an efficient way of conserving spectrum space, in which a large number of closely- and evenly-spaced carriers, individually and separately modulated with a sub-band of frequencies and/or phases (e.g. Figure 11.10) are broadcast in a 'slot' in the frequency spectrum. The perfectly even distribution of power and energy in the channel is illustrated in Figure 11.11, and compares very favourably with that shown in Figures 3.13 and 5.2 for conventional analogue AM TV broadcasts. OFDM and 64-QAM permits a transmission rate of 60 Mbits/second in a standard 8 MHz-wide TV channel, sufficient for several high-quality TV channels or many more standard-quality ones.

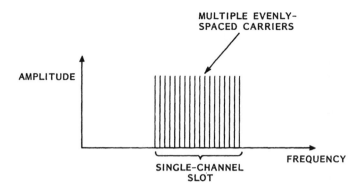

Figure 11.11 *OFDM carriers efficiently utilise spectrum space. Each carrier conveys a fragment of the total data*

Because of the even power distribution throughout the channel-slot used for OFDM transmissions, and the relative ruggedness and noise-immunity of the system (perfect pictures can be obtained with carrier-to-noise ratios of less than 20 dB in a practical system) the carrier power required for transmission is very low compared with that for ordinary terrestrial TV channels. The risk of interference in co- and adjacent channels is reduced accordingly, opening the way to the use of channels which would be 'taboo' for conventional broadcasts. Coverage of area and population is increased by the use of digital TV transmission, then.

Quality Graduations

In a TV broadcasting system where digital transmissions are the sole carriers, the transmission format must take into account the widely-differing requirements of everything from widescreen high-definition TV through conventional receivers to VTRs and small-screen portable and pocket TV sets, the latter of which do not need, and cannot use, the full definition and quality available, and for which smaller, simpler and less expensive decoders are appropriate. With suitable coding and multiplexing systems this can be achieved as Figure 11.12 shows. In the case of the VTR, a digital type is fed via a transcoder/interface to prevent the picture quality degradiation associated with digital decoding to baseband and subsequent reconversion to the digital format used for recording onto tape.

One of the advantages of analogue transmissions and signal paths is that picture quality deteriorates gradually as reception conditions gradually degrade; as the signal-to-noise (S/N) ratio increases, for instance with increasing distance from the transmitter, the picture becomes more snowy, but remains watchable until reception conditions are very poor. The characteristic of a conventional digital signal, on the other hand, is that picture quality remains good until a certain carrier-to-noise (C/N) ratio is reached, after which it deteriorates very rapidly. The actual C/N ratio which can be tolerated depends on the amount of error-coding provided with the signal, or its *redundancy ratio*.

157

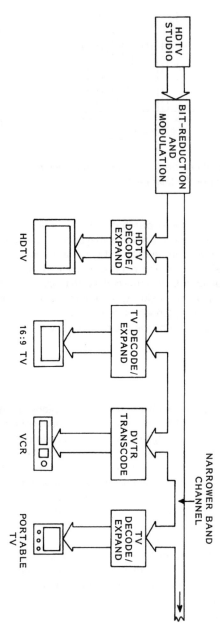

Figure 11.12 *Multi-standard digital highway for simultaneous signal distribution to various standards*

There are two techniques available to overcome this problem, offering what has become known as *graceful degradation*. One is to provide more error protection for data representing the components of the picture which go to make up a relatively low definition image (such as those which would be used in small-screen and portable TV applications) so that higher-order sets could revert to them, in one or two stages, as reception deteriorates. The second is to provide error concealment in the receiver's decoder, which increases its cost but provides the incidental advantage of improving picture quality in the presence of minor errors in reception. Here the decoder has the facility to discard complete blocks of coded picture samples and replace them with new blocks corresponding to an intelligent guess of what they might have contained, typically based on samples from the previous picture frame.

In practice this technique is very successful due to the complex algorithms which have been developed for it. Figure 11.13 illustrates in graphical form the principle of graceful degradation, along with the characteristics of analogue and conventional digital broadcasts.

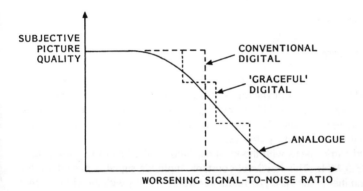

Figure 11.13 *The principle of graceful degradation in digital TV transmission*

12 TV sound systems

In Chapters 5 and 8 we briefly touched on sound systems for TV. Here the subject will be examined more closely, with particular regard to the digital (Nicam for terrestrial and MAC for satellite) and multi-carrier (analogue satellite) systems currently used for entertainment broadcasts. First let's explore in a little more detail the conventional FM sound system which has served us for many years and continues to do so.

FM mono system

The monaural sound is transmitted on its own frequency-modulated RF carrier with ±50 kHz maximum deviation, at a level 10 dB below that of the vision carrier. In European systems B/G the spacing is 5.5 MHz above the vision carrier. For the UK System I scheme it's 6 MHz above the vision carrier frequency, and we'll use that for descriptive purposes.

The sound carrier beats with the local oscillator in the tuner to produce an IF output at 33.5 MHz. This low-level constant-amplitude signal is passed via the vision IF amplifier to the vision demodulator where it beats with the vision carrier to produce a 6 MHz signal, still frequency-modulated, which can then be selected, amplified and demodulated. This *intercarrier* system has several advantages: tuning errors and drift have no effect on the carrier frequency, which is governed solely by the very accurately-maintained vision–sound spacing at the transmitter; the sound carrier benefits from the gain provided by the vision IF amplifier; and the sound processing circuit is a simple one.

Receiver circuit

Figure 12.1 shows a typical intercarrier sound system, contained entirely in one IC, for a basic TV set. The wanted 6 MHz carrier is selected from the demodulated vision carrier by a ceramic filter, resonant at 6 MHz and having a bandwidth of about 200 kHz. The separated sound carrier passes into the IC at pin 1 for passage through several stages of amplification with limiting, which latter clips off any amplitude modulation caused by the vision signal: the AM rejection is about 55 dB. This clipping action produces a squarewave output, rich in harmonics of the 6 MHz carrier frequency. They are suppressed in a low-pass filter which

Figure 12.1 *FM receiver and audio circuit for terrestrial monaural sound*

restores the carrier waveshape to approximately a sinewave for application to the FM demodulator of the quadrature synchronous type. It works on the sample-and-hold principle which we've met before in these pages. The sampling gate is opened for an iinstant during each carrier cycle at a time when an unmodulated carrier wave is passing through the zero point, so the sample taken corresponds to zero volts. As the phase/timing of the carrier signal advances and retreats in sympathy with the modulating sound signal, the sampling action produces an output proportional to the frequency deviation, an exact replica (after de-emphasis) of the original sound.

A reference carrier is required to produce the demodulator gate pulses. It's generated by the high-Q tuned circuit L1/C3 whose flywheel effect averages the carrier frequency, thus providing a constant-phase, 90° reference feed. In current practice this tuned circuit takes the form of a ceramic filter.

Next comes a voltage-controlled amplifier (VCA) whose gain depends on the DC voltage present at IC pin 6: the reproduced sound level is proportional to the applied voltage, which is produced by the control system as described in Chapter 8. In many modern sound processors ICs the control voltage is internally derived from data on the serial control bus as illustrated in Figure 8.10.

The demodulated audio signal is de-emphasised next, by an RC network whose C element is external to the IC at pin 12. The audio signal is now ready for application to the driver and output stages which (in the simple system shown here) is inside the same IC package. The output stage generally consists of a push–pull pair of transistors operating in Class B mode, the DC mid-point voltage being isolated from the loudspeaker by coupling capacitor C10. The value of the resistors connected to pin 7 of the IC determines, as part of a negative feedback loop, the AC gain of the output stage, the capacitors connected to pin 8 setting the amplifier's upper frequency limit. R5 and C9 form a Boucherot cell which suppresses any tendency for HF oscillation due to the inductive load of the loudspeaker and its wiring.

A small IC of the type illustrated here can provide an audio output power of about 3 W, depending on the supply voltage and the heatsink capacity (if any) provided. This type of IC, or a small power amplifier chip fed from a separate intercarrier sound amplifier/demodulator IC, easily caters for the needs of a portable TV or an economy large-screen TV using a single loudspeaker. Although for the purpose of description here we've examined a dedicated sound channel IC, in many sets the intercarrier sound and audio preamplifier stages are incorporated in a more complex IC package which performs many other functions.

Stereo sound

Most large-screen TV sets are stereo types, fitted with two (or two sets of) loudspeakers, and relatively high power amplifiers to drive them. Balance, volume and tone controls are provided by the microprocessor control system, often via the two-wire control bus described in Chapter 8. Some sets have provision for processing a monaural sound signal to give it a 'spacial' or pseudo-stereo effect by differentially manipulating the phase or amplitude of the signals in the left and right amplifier channels. Circuits have also been devised to emphasise the difference between the L and R audio signals from a stereo source to artificially widen the perceived sound field: the loudspeakers of a stereo TV set are necessarily close together, and this 'stereo-wide' system partially compensates for this.

Stereo TV sound can come from several sources: Nicam and MAC broadcast transmissions; stereo VTRs; video discs; camcorders; and the FM stereo sound carriers of satellite broadcasts, which we shall now examine.

Satellite FM sound

With most conventional satellite TV transmissions using FM picture modulation (more of that in Chapter 13) there are FM carriers for the sound, very similar to those used for terrestrial mono TV sound. The main difference is that there are more of them! Figure 12.2 shows a typical satellite TV channel spectrum, with five sound carriers on the HF side of the baseband video signal, at 6.5, 7.02, 7.20, 7.38 and 7.56 MHz. The main carrier at +6.5 MHz is used for the monaural sound signal, with a bandwidth of 20 Hz to 15 kHz. Carrier deviation is ±85 kHz and the pre-emphasis time-constant 50 μs.

Thus the FM signal processing system is exactly the same as that already described for terrestrial TV sound transmissions save for the operating frequency – in the example just quoted the filter and demodulator are tuned to 6.5 MHz. The simplest satellite TV receivers demodulate only the main sound carrier, providing a single audio output for coupling to a TV set or VCR.

Auxiliary carriers

The auxiliary sound carriers have only ±50KHz maximum deviation, though the audio bandwidth is the same, resulting in a lower modulation index. To prevent this from impairing the S/N ratio a noise-reduction system is used – with Astra transmissions the *Wegener Panda 1* type is employed. The term Panda is derived from 'processed narrow-deviation audio', a form of adaptive pre-emphasis. The dynamic range of the audio signal is compressed before transmission so that, in relative terms, high-level signals are attenuated and low-level ones boosted, the amount of compression also being frequency-dependent. The inverse of this process is applied at the receiver so that the dynamic range of the signal is restored. In the process the noise component is suppressed.

The system has much in common with the companding principle used in hi-fi VCRs (Chapter 20 and Figure 20.7) and with Dolby noise-reduction techniques. It's implemented by IC-based VCAs whose control voltages are derived from the signal itself via filters. Without noise-reduction the signal-to-noise ratio of a narrowband satellite TV sound channel is about 50 dB: the Wegener Panda 1 system provides an improvement of about 18 dB so that the subjective S/N ratio approaches 70 dB, which is very good for an analogue transmission. Table 12.1 shows the uses to which the auxiliary carriers are put, for stereo and multi-lingual sound in a typical satellite broadcast channel.

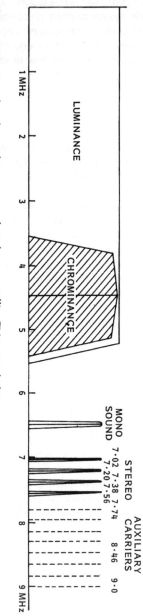

Figure 12.2 *Baseband signal spectrum of analogue satellite TV transmission*

Table 12.1 *Typical assignments of audio carriers in a satellite TV transmission*

Mode	7.02 MHz	7.2 MHz	7.38 MHz	7.56 MHz
1	language 1L	language 1R	language 2L	language 2R
2	language 1L	language 1R	language 2 mono	language 3 mono
3	language 1 mono	language 2 mono	language 3 mono	language 4 mono

Carrier conversion and selection

It is impractical to provide separate filters and demodulators for all the spot-frequency sound carriers shown in Figure 12.2, especially as a maximum of two (e.g. L/R stereo) are in use by the viewer at any one time. Instead, a superhet configuration is used to convert the wanted carriers to fixed frequencies which can be dealt with by a single pair of fixed-tuned filters and demodulators. This technique has the advantage of versatility: it works with any pair of carrier frequencies, including the piggy-back radio stations (Figure 12.2) supported by many transponders on carrier frequencies between 7.74 and 9 MHz, while in sophisticated systems the required sound channel can be user-programmed per transponder and stored in memory.

There are various ways of implementing the sound-superhet technique. A common one is illustrated in block-diagram form in Figure 12.3, where two fixed-frequency oscillators run at 17.72 and 18.08 MHz. The output from one or the other is selected and fed to one of the gates of the dual-gate FET (field effect transistor) mixer, where it beats with the incoming signal to produce IFs at 10.52 MHz (R channel) and 10.7 MHz (L channel). With the 17.72 MHz oscillator switched in, the R sound comes from the 7.2 MHz carrier (17.72−7.2=10.52 MHz) while the L sound comes from the 7.02 MHz carrier (17.72−7.02=10.7 MHz). When the control system selects the output from the 18.8 MHz oscillator, the R and L sound signals come from the 7.56 MHz and 7.38 MHz carriers respectively.

Instead of having fixed oscillators a variable or programmable oscillator can be used: by varying its frequency it's possible to tune – like a radio receiver – through the band of sound carriers associated with each satellite broadcasting channel. Figure 12.4 shows how this may be arranged. The variable-frequency local oscillator is controlled by an IC containing a programmable divider and a 4 MHz crystal-controlled reference oscillator. These form a frequency-synthesis tuning system which is governed by the receiver's microprocessor control IC. The oscillator's output is fed to gate 1 of the mixer FET, whose other gate receives the whole spectrum of sound carriers from bandpass filter FL1, whose passband extends from 5 to 9 MHz.

The mixer's output contains all the sound carriers in the form of beat products in the range 8–13 MHz. Those centred on 10.52 and 10.7 MHz are selected by ceramic filters FL2/FL3 and FL4/FL5 respectively and applied to pins 11 and 14 of the dual FM demodulator IC2. By programming IC1 any pair of sound carriers 180 kHz apart (see Table 12.1) can be brought into line with the filters and demodu-

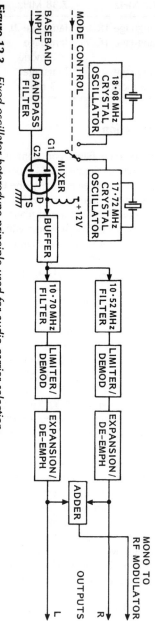

Figure 12.3 *Fixed-oscillator heterodyne principle used for audio carrier selection*

Figure 12.4 *'Tuneable' audio carrier selection principle based on EEPROM memory in the control system*

lators. In a typical design the sound is tunable from 5 MHz to 9 MHz in 10 kHz steps, and any required point can be stored in non-volatile memory in a tuning/control system similar to that described in Chapter 8. The L and R outputs pass from the dual-demodulator IC to de-emphasis circuits which can be switched between simple linear (50 µs) or Wegener-Panda expander operation, the latter involving a purpose-designed and licensed-use IC. It's interesting to note that in this mode of audio operation the sound carriers undergo a quadruple-superhet process: their frequencies are changed at the dish-mounted LNB (low-noise block), at the indoor tuner (both to be described in the next chapter) at the vision demodulator and at the sound carrier frequency changer.

Digital TV sound systems

Nicam (terrestrial) and MAC (satellite) sound systems are very similar in their modes of operation, and the overview which follows, although based on Nicam, is equally applicable to the audio component of a MAC broadcast except in respect of the modulation and encoding of the digital data.

The Nicam system conveys audio data in phase-modulation of a low-power carrier placed just above the conventional FM sound carrier of a TV broadcast. It has capacity for two totally independent sound channels, which are generally the L and R components of a stereo signal, but can be used for bi-lingual sound tracks or data for (e.g.) downloading to a computer.

Figure 12.5 outlines the processes involved in Nicam transmission. The studio L/R sound feeds are first pre-emphasised and passed to an analogue-to-digital (A/D) converter with a clocking rate of 32 kHz, which means that samples are taken at intervals of 31.25 µs. To prevent aliasing the upper frequency limit of the L and R audio input signals is restricted by filters to 15 kHz. Each of the channels is simultaneously sampled to 14 bits to give 16 238 possible levels, after which the bit count is reduced to ten to conserve transmission bandwidth.

The 14 to 10-bit conversion process is a complex one, designed to retain most of the advantages of 14-bit processing – it's illustrated in Figure 12.6. From top to bottom of the diagram every possible bit combination for a 14-bit word is shown: where the bit value is irrelevant it is shown as an X. The shaded areas indicate the bits removed for transmission. The most significant bit (MSB) passes through regardless, but the next bits are deleted if they are the same as the MSB, up to a maximum of four. If this process leaves more than ten bits, sufficient bits are trimmed from the LSB (least significant bit) end to reduce the length of the word to ten. Thus is the digital signal compressed, and as long as the decoder (at the receiving end) knows what the compressor is doing from moment to moment it can expand the data back to 14-bit form.

A careful look at Figure 12.6 shows that 10-bit resolution is given to large signals (coding range 1), 11-bit resolution applies in range 2, 12-bit in range 3 and so on up to the 14-bit range 5 which represents the quietest and most vulnerable sounds. This 'progressive coding' system is very effective in terms of bandwidth economy, while losing little in comparison to a linear coding system. Ordinarily a 10-bit data word offers an S/N ratio of about 60 dB in the reproduced signal. With the companding system described here the S/N ratio approaches the 84 dB theoretically available from a 14-bit system.

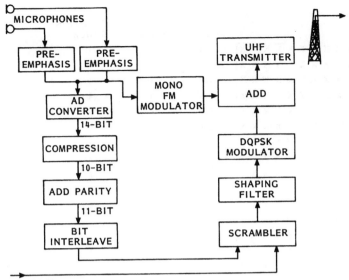

Figure 12.5 *Block diagram of the processes involved in the transmission of a Nicam stereo sound signal*

Protection and scale factor

A *parity* bit is added at the end of each 10-bit word to provide a check on its validity at the decoder. Here even parity is used so that the first six bits of the received word, when added to the parity bit, give a '0' result. The parity bits are modified according to a look-up table held in ROM at each end of the chain to signal the scaling factor in a three-bit word shown right of centre in Figure 12.6. If the missing bits are on the left-hand side of the diagram the scale factor data enables them to be recreated at the decoder; if, however, they are on the right of the block they are lost forever. The words of which these latter form a part are not crucial ones in terms of noise, though, because they represent loud sounds or transitions between two widely-different levels, where the noise is masked.

For scale-factor signalling purposes the 11-bit (10 plus one parity bit) words are grouped together in blocks of 32, each block occupying 1 millisecond. Since only one scale-factor message is sent for 32 consecutive words, some of them do not get optimum expansion accuracy. In fact the scale-factor conveys data on the magnitude of the largest sample in each block. While this inaccuracy increases the quantisation noise in the sound signal, it only happens where the 'busyness' of the signal itself disguises or masks it. The scale-factor data is sent frequently enough to track the fastest perceptible loudness transitions, and that is the key to subjective noise reduction.

The parity system gives good protection against corruption of the data words for single-bit errors, but impulsive interference can take out complete data words, making necessary further protection. It's achieved by a simple interleaving system in which the data is written into memory and then read out non-sequentially

	SCALE FACTOR	CODING RANGE	PROTECTION RANGE
0 1 X X X X X X X X X X X X	111	1	1
0 0 1 X X X X X X X X X X X	110	2	2
0 0 0 1 X X X X X X X X X X	101	3	3
0 0 0 0 1 X X X X X X X X X	011	4	4
0 0 0 0 0 1 X X X X X X X X	100	5	5
0 0 0 0 0 0 X X X X X X X X	010	5	6
0 0 0 0 0 0 0 X X X X X X X		5	7
0 0 0 0 0 0 0 X X X X X X X	000	5	7
1 1 1 1 1 1 1 X X X X X X X	OR 001	5	7
1 1 1 1 1 1 1 X X X X X X X		5	7
1 1 1 1 1 1 0 X X X X X X X	010	5	6
1 1 1 1 1 0 X X X X X X X X	100	5	5
1 1 1 1 0 X X X X X X X X X	011	4	4
1 1 1 0 X X X X X X X X X X	101	3	3
1 1 0 X X X X X X X X X X X	110	2	2
1 0 X X X X X X X X X X X X	111	1	1

X = IRRELEVANT BIT   ▨ = BITS DELETED FOR COMPRESSION

Figure 12.6 *Nicam coding/compression table showing how the 14-bit words are reduced to 10 bits for transmission. Bits in the shaded areas are not transmitted*

according to an address sequencer held in ROM. The data readout order ensures that bits which were originally adjacent become separated by at least fifteen other bits. An error burst in the received data can corrupt several consecutive bits, but when the words are reassembled at the receiver (by a memory-plus ROM system reciprocating that at the encoder) the errors are distributed among several of them so that the damage to each is usually minor, and capable of repair by parity correction and/or error concealment as necessary, techniques which are universally used with PCM sound systems. The digital signal now has enough protection to enable it to pass unscathed through all but the worst propagation conditions.

Sound/data block

Along with the audio data it's necessary to send 'housekeeping' data to control the decoding process. Figure 12.7 shows the composition of a single stereo broadcast data frame, each of which spans 1 millisecond and contains 728 bits. The frames are sent end-to-end with no gaps between them.

The block starts with a frame alignment word (FAW) which synchronises the decoder at the receiver. It's the same in every frame, and consists of the eight-bit sequence 01001110. Next come five control bits C to define the type (stereo, bilingual, data) of signal being sent and thus to control the decoder's operating mode and the switching and routing of its output signals. The next data block AD is reserved for 'additional data' and can be used for various purposes.

Figure 12.7 *Nicam data frame as transmitted for a stereo sound signal*

The initial section of the frame has used 24 bits. The remaining 704 convey the stereo or dual-channel sound data in 64 11-bit words, the A channel (stereo left) and B channel (stereo right) samples being sent alternately throughout the period, 32 of each. The two sound channels (plus the control/data preamble) are thus transmitted as a single serial data stream, and the bit rate of each is approximately doubled as a result. This is a form of time-division multiplex (TDM) and the required time compression is achieved by writing each set of data, A and B, simultaneously into memory then reading them out alternately at double speed. TDM is used in many data storage and transmission systems to match signal density to channel bandwidth. The order of A and B samples shown in Figure 12.7 represents the sequence before bit interleaving, described above, takes place. Only the 704 bits of sound data are interleaved.

Data scrambling

The Nicam data stream contains certain fixed patterns. If it is used as it stands to modulate a carrier, the sidebands produced will contain fixed patterns likewise, and that is undesirable from the point of view of interference to other users of the broadcast band, so the sideband energy is evened out by 'scrambling' the data. It's not possible to scramble the data bits in a completely random way because they could not be recovered at the receiver. A simple circuit which generates binary digits in what appears to be a random way but whose output is predictable and repeatable is called a pseudo-random sequence generator (PRSG). Identical PRSGs are used at both ends of the chain, and run in synchronism throughout the data frame, starting at the end of the FAW. Their outputs are added modulo-two (in an exclusive-OR gate) to the data. Thus the data are scrambled for transmission, giving the effect of a completely random bit-stream, then descrambled at the decoder.

Modulation and transmission

A relatively simple carrier modulation system is used for Nicam; it's called digital quadrature phase shift keying, DQPSK for short. The DQPSK carrier is constant in frequency, but has four possible phases. At the transmitter a serial to two-bit parallel converter changes the serial data into a series of two-bit pairs which can take one of four forms: 00, 01, 10 and 11. Each of these moves the carrier phase by a different amount, as shown in Table 12.2, from its previous state. The effect is shown in the phasor diagram of Figure 12.8. Carrier phase changes thus

171

take place at two-bit intervals with a maximum shift of 180° or half a cycle of carrier frequency.

Table 12.2 *Nicam carrier phase changes for all possible combinations of two-bit data pairs*

Input data	Carrier phase
00	0°
01	−90°
10	−270°
11	−180°

In practice the phase does not jump from one state to another, but moves smoothly due to a low-pass filter in the data path to the modulator.

Figure 12.9 shows the spectrum of a terrestrial TV broadcast with mono FM and Nicam carriers at the HF end of the channel slot. The Nicam carrier, at +6.552 MHz is broadcast at the low level of −20 dB with respect to the vision carrier, representing a *power* ratio (peak vision carrier to Nicam carrier) of 100:1.

Nicam reception

The phase-modulated Nicam carrier is tuned at the TV receiver or VTR as part of the sound/vision 'bundle'. It beats with the local oscillator inside the tuner to produce its own IF signal, still phase-modulated, which is allowed to beat with the vision carrier (just like the 6 MHz mono sound carrier) to generate a difference signal at 6.552 MHz for the UK system I transmissions. A diagram of a typical set-up for Nicam reception and carrier-handling is given in Figure 12.10.

Demodulation and decoding

The Nicam decoder's operation, in grossly simplified form, is conveyed in Figure 12.11. The four-phase modulated carrier at 6.552 MHz is intercepted by a filter and passed into the Nicam IC, where it first encounters an AGC control stage to iron out level fluctuations. Next comes a phase detector, timed by a synchronised crystal oscillator running at the carrier frequency of 6.552 MHz. The phase detector's outputs pass into a pair of adaptive slicers, which optimise the points at which the signal is sampled in the following stage, the QSPK detector, gated by switching pulses from a data-rate clock running at 5.824 MHz, again governed by a quartz crystal.

Now the signal has been demodulated into bit-pairs, and the next stage is to convert them to a serial datastream: the parallel to serial (P/S) converter is clocked by the 5.284 MHz oscillator mentioned above, which is synchronised to incoming data by its own phase-locked loop.

The demodulated data now undergo decoding, which is the inverse of the processes described earlier; the operation is illustrated on the right-hand side of Figure 12.11. The decoding is synchronised by the FAW and governed by the

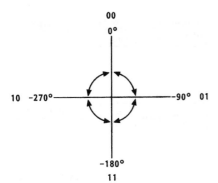

Figure 12.8 *Carrier phase shifts for the four bit combinations of Table 12.2*

Figure 12.9 *The position of the Nicam carrier in the demodulated signal spectrum*

control bits. The local PRSG is reset at the end of the FAW sequence and added to the data to descramble it. De-interleaving is carried out with the help of the on-board RAM (typically 3 of 64 × 11-bit) whose readout, in the order dictated by the built-in 'codebook' ROM, re-creates the original 11-bit data words for passage to the 14-bit expander stage, whose operation depends on the scale factor shown in Figure 12.6 and the parity bits which convey it. The parity bits are also used here in the processes of reconstruction, by repair or discard-and-interpolate, of any damaged or corrupt data. Data for L and R channels are produced alternately and in rapid succession.

The final process inside the Nicam IC is D/A conversion, in which the data carried by each 14-bit word are translated into a voltage level, generally by charging a capacitor at constant current for a period dependent on the data contained in the word. Separate sampling capacitors are provided for L and R channels, synchronously selected by the L/R ident switching signal derived from the system clock. Emerging from the IC, then, are two sets of step-waveforms representing an analogy of the original L and R audio waveforms.

Figure 12.10 *Derivation of the Nicam carrier signal in a 'parallel' IF system. Some receiver designs use a single-output SAW filter and extract the Nicam carrier from the vision demodulator via a 6.552 MHz filter*

Post decoder

The 'raw' L and R signals are filtered in low-pass circuits (ft=15 kHz) to smooth them and eliminate the discrete steps they contain. Finally de-emphasis is applied in a frequency-selective feedback circuit to restore tonal balance and suppress noise. Now back in 'hi-fi' form, the stereo signals undergo a switching/routing/selection process before application to their amplifiers and loudspeakers, or to futher processing for recording on tape.

Surround sound

Home stereo sound sources, including commercially-produced video cassettes, Nicam and analogue-FM transmission systems, are capable of carrying *surround-sound*, encoded on a plan devised by Dolby Labs. The system, originally designed for use in cinemas, is now available to the 'home cinema' enthusiast in commercially-available decoders, available in black-box form or incorporated into TV sets, VTRs and satellite receivers.

The surround sound 'package' is carried solely by the two (L and R) stereo channels in the broadcast or storage medium, but does depend on them having well matched phase and amplitude responses. Figure 12.12 shows how the Dolby surround encoder works. It accepts four separate input signals, left, centre, right, and surround (L, C, R, S) and creates from them two outputs: left-total and right-total – Lt and Rt. The L and R inputs pass straight to the Lt and Rt outputs without modification. The C input is split equally between the Lt and Rt channels with a 3 dB level reduction to maintain the correct proportion of acoustic power in the mix. The S input (carrying the 'surround' channel effects) is also divided equally between Lt and Rt channels, but it first undergoes three processing steps. Initially it's bandwidth-limited in a filter which only passes frequencies in the range 100 Hz to 7 kHz. Next a form of noise reduction is applied in a modified Dolby-B NR

Figure 12.11 *Single-chip Nicam decoder incorporating demodulator, decoder and D/A convertor*

175

encoder. Finally the C-signal undergoes plus and minus 90° phase shifts to create a 180° phase differential between the signal components added to Lt and Rt.

The surround-encoded signal now passes over the broadcast medium or into the record system. When reproduced by ordinary stereo equipment there is no loss of compatibility; the full separation between L and R signals is maintained. Neither is there any theoretical loss of separation between the centre and surround signals: since the surround signal is recovered by taking the difference between Lt and Rt, the (identical) centre channel signals in Lt and Rt cancel out in the surround output. Similarly, since the centre channel is derived from the sum of Lt and Rt, the equal and opposite surround channel components cancel out in the centre output. Maintenance of low crosstalk between centre and surround signals requires close matching between the two transmission/storage channels. If a mismatch between them causes the centre channel components in (e.g) Lt to differ from those in Rt, centre information appears in the surround channel as unwanted crosstalk.

Pro-Logic decoder

There are two forms of surround decoder, the passive and the Pro-Logic types. Of these the latter is most popular and effective; an outline of its operation is given in Figure 12.13. The Lt and Rt signals pass relatively unmodified through the decoder to form the main L and R outputs. Lt and Rt also pass through a Pro-Logic adaptive matrix, shown at the bottom of Figure 12.13. Here they are separately bandpass filtered to 100 Hz – 13 kHz before being fed to four full-wave rectifiers, two directly, one via an L/R adder and one via an L/R subtractor. The DC control voltages thus derived are processed in logarithmic-difference amplifiers whose outputs indicate the *dominance vector* in the soundfield, regardless of absolute signal levels in the channels. It is used to apply directional separation enhancement to a degree governed by the dominance in the transmitted signal, and works not only in the four directions represented by L, R, C and S, but at any point in the 360° surround-sound field in which the viewer sits.

The two control voltages are bipolar, and represent, respectively, dominance along the left/right axis and along the centre/surround axis. Each control voltage is

Figure 12.12 *Principle of operation of a Dolby surround encoder*

Figure 12.13 *Pro-logic surround decoder. The upper part of the diagram shows the decoder as a whole, while the lower section enlarges on the adaptive matrix*

177

continuously monitored by a threshold switch which – once the relative dominance exceeds its trigger level – switches the circuit to a fast mode of operation, quickening its response to change in level. A polarity splitter resolves the two bipolar dominance signals into four unipolar control voltages E_L, E_R, E_C and E_S which represent the soundtrack dominance in electrical terms embodying psychoacoustic properties. They are applied to the signal-cancelling VCA (voltage-controlled attenuator) stages. With two input signals (Lt and Rt) and four control voltages, the eight VCAs can generate eight variable sub-terms; with the Lt and Rt inputs, ten individual terms are available. To construct a decoded output signal, portions of each of these ten terms are selectively added or subtracted with a predetermined weighting factor in the combining networks. Selection of the appropriate magnitudes and polarities for the 40 summed components gives directional enhancement and non-dominant signal redistribution to the reproduced soundfield while maintaining constant accoustic power for the signal components.

Returning now to the top section of the decoder block diagram in Figure 12.13, the surround signal passes through an anti-alias filter, rolling off at about 7 kHz. Next it encounters a variable time delay, in practice a RAM store, whose delay period (typically 20 milliseconds) can be set by the user to match the propagation time of the sound from the rear (surround) loudspeakers to the listening position. The idea is not to create a 'spacious' effect by means of an echo, but to improve the sense of clarity and direction of *front* channel sounds. It is based on the *Haas* or precedence effect. Emerging from the delay stage the S signal is again filtered to 7 kHz before passage to the noise-reduction decoder, which applies a reciprocal effect to that operating at the sending end. Finally the S signal, along with the others, passes through a VCA master volume control on its way to the power amplifiers which drive the loudspeakers.

In practice two loudspeakers are generally used at the rear of the listening room to provide the surround sound, while the centre speaker is placed very close to the viewing screen, primarily to reproduce dialogue, which needs to be tied closely to the picture, direction-wise. Low frequency sound has no perceptible directional characteristic, and can be reproduced by a single woofer unit placed almost anywhere in the room.

The noise sequencer shown in Figure 12.13 is used for setting up and matching to room-size and accoustics in an inbuilt computer-controlled installation program. The entire Pro-Logic decoder is embodied in a single IC, typically with an outboard RAM to provide the delay for the surround channel. Several operating SFP (soundfield processing) modes are available, examples of which are Church, Cinema, Stadium, Disco and Concert Hall. They alter echo and reverberation characteristics, and can be stored in EEPROM (user's software) and called up at will by the remote control keys. They are conveyed to the surround decoder chip by a two-wire serial control bus like that described in the 'control systems' section of Chapter 8.

13 Satellite TV

A relative newcomer to the broadcasting field is the practical realisation of space transmitters. Before we attempt to describe them in any detail, we need to understand a little of how a satellite is positioned and controlled so as to appear stationary to an earth-mounted receiving installation.

As we are well aware, the force of gravity is a dominant one on earth, tending to pull loose objects towards the centre of the earth. There is another effect called centrifugal force which operates on spinning objects, this time pulling outwards, away from the centre of the spin-arc, as demonstrated by the domestic spin-drier, certain fairground rides and a conker on a string. The strength of centrifugal force depends entirely on the orbital speed of the object in question. For any given orbital speed, then, there must be a point in space where centrifugal force acting on an orbiting satellite just cancels the gravitational pull of the earth, so that these balanced forces hold the satellite at a fixed distance above the earth's surface without any need for motive power from the satellite itself. If we require the satellite to appear to stand still relative to Earth, it will need to orbit at exactly the same speed as Earth itself – once every twenty-four hours. With the orbital speed thus rigidly fixed, the distance from Earth for the satellite comes out at 35 800 km (22 200 miles) to satisfy the gravity versus centrifugal force equation. So we have two fixed parameters for our *geostationary* satellite, orbital speed (which comes out at 11 069 km/hour for one revolution per day) and the altitude of 35 800km. It must necessarily be over the equator to prevent an apparent daily North–South oscillation. When all these parameters are satisfied, the satellite will effectively hang stationary in the sky. From time to time drift will inevitably take place of its position and attitude, and this is corrected by small on-board jet motors which operate on a *servo* principle to stabilise and maintain the satellite's attutude to earth.

Thus we can set up a transmitter in the sky, but at first sight its position above the equator is bad news for the UK, situated as it is some way into the Northern Hemisphere. Plainly, a radio beam from the satellite will come in to us at an oblique angle rather that directly downwards; in the UK the elevation angle from the horizon varies from 17° in the Shetlands to almost 28° in West Cornwall. Except in built-up areas, a clear line-of-sight to the sky can usually be secured at this angle from ground level, so that receiving aerials can be mounted at ground level in many domestic situations, perhaps in a back or side

Figure 13.1 *Typical signal footprint for a pan-European satellite. This one shows that of the Astra 1A beam*

garden; only in the shadow of buildings or other obstructions will it be necesary to mount the receiving dish on a wall, roof or tower. In urban areas where these problems exist they can be solved by the use of a 'community dish' with local cable distribution to individual homes.

The slant 'radio-illumination' effect stemming from the satellite's position over the equator means that a narrow transmission beamwidth can cover the entire UK; for a *footprint* like that shown in Figure 13.1 the beam need only be 1.8° in the 'vertical' plane and 0.7° across. Because all the transmission energy can be concentrated in this very narrow beam by highly directional (almost like a gun!) transmitting aerials at the satellite, the use of radio energy is very efficient indeed, though the narrow beam calls for very great accuracy in the alignment of the satellite in space, and the receiving dish on the wall! They have to be held within ±0.1° and ±0.5° of nominal pointing position respectively.

To give an impression of the radio-torchbeam-from-space effect, and the reason for the elliptic/potato shape of the footprint, Figure 13.2 represents the relative positions of the satellite, the equator, and the UK.

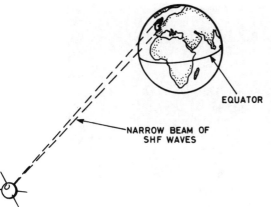

EQUATOR

NARROW BEAM OF
SHF WAVES

SATELLITE

Figure 13.2 *The relationship between the Earth, equator and a geostationary satellite*

Satellite power

In Chapter 5 we discussed the effect of a directional radiating aerial on the apparent power of the transmitter, and saw that ERP was the product of true RF power multiplied by the 'gain' of the transmitting aerial. The very narrow beam from a satellite (in the UK case, 1.8° × 0.7°) concentrates the power beautifully, and from an RF output power of about 150 W we can realise an ERP of 3000 kW! The only type of radiating aerial capable of launching such a narrow RF beam is a parabolic dish, and this is the basis of the satellite's transmitting element. It may be between 1 and 3 m in diameter, depending on the beam angle required. The true RF power of the satellite contrasts remarkably with that required for terrestrial broadcasts – on the one hand 50 W for good coverage of the whole of the UK, on the other many many megawatts, indifferent coverage of some areas and 1000 individual transmitters to provide and maintain, in each case for one TV channel.

The power to run the transmitter and the peripheral equipment in the satellite itself comes from banks of solar cells drawing energy from the sun. These are mounted on 'wings' springing from the satellite body, and are servo-controlled to keep them at right-angles to the sun's rays. This accounts for the characteristic dragonfly shape of a broadcast satellite: see Figure 13.3.

The DC power provided by the solar cells needs to be about 2 kW; much depends on the efficiency of the transmitters and the requirement of other, 'housekeeping' equipment on board. Some countries, due to their size and position relative to the equator, require greater beamwidth than others, and for larger beamwidths the RF power required increases rapidly. As an example, if the UK were directly below a satellite on the equator we would need a fivefold increase in power to achieve the same field strength over the same area.

Figure 13.3 *A typical TV broadcast satellite*

The satellite as an 'active reflector'

Of course, the programmes are not originated on board the spacecraft, so it must also be fitted with a receiving aerial and the necessary transposing equipment. Coming up to it from earth will be the programme material and signal routing instructions, along with engineering control signals to cater for 'housekeeping' and positional control. This uplink also takes place in the SHF band with a narrowbeam signal between parabolic dishes. Slotted into the satellite's output signals are data streams containing information about the internal workings of the spacecraft and its motor, power and signal-processing sections for monitoring on Earth. The up- and down-links are shown in Figure 13.4.

The active life of the craft comes to an end when the fuel for the correction motors runs out, assuming that no catastrophe occurs to the other systems to disable the satellite. As in all space-equipment design, crucial components are duplicated, and a great deal of potential redundancy is designed into every part of the vehicle. At the end, the satellite can be 'kicked off' into obscurity in deep space by a final burst of the positional motors at the very last, to make room for its successor!

Satellite signal reception

Leaving aside now the 'Star Wars' atmosphere of the orbiting satellite, it is time to return to the back garden to see what will be required in the way of receiving aerials for space transmissions at 11 GHz. While in theory a dipole with reflector and directors might work, the reality is that at the wavelength of a 11 GHz carrier (about 2.5 cm) the conventional aerial would be small enough to slip into a shirt

Figure 13.4 *Arrangement of up- and down-links for the DBS satellite*

pocket and would certainly not have enough 'meat' in it to extract a usable (or even detectable!) signal. We need to capture the beam over a relatively wide area and focus it on a dipole to achieve sufficient field strength to make a usable signal. For this a parabolic dish is called for, with some form of dipole at its focal point.

The larger the dish, the more RF energy it can intercept and concentrate on the pickup element; thus dish size has a counterpart in element-count in a conventional Yagi array. Again, the question of *phase* is relevant because all the reflected cycles of RF carrier must arrive in phase at the dipole in order to reinforce each other; see Figure 13.5. If any arrive out of phase a cancellation effect will take place with consequent loss of gain. This requires tremendous accuracy in the manufacture of the dish which must not deviate from a true parabolic shape, and must have a good surface finish. Any physical damage, warping or contamination of the dish surface will reduce the efficiency of the receiving aerial, which in good conditions should be about 60 per cent.

Figure 13.5 *Parabolic receiving dish showing incoming parallel RF beam and the in-phase reflection to the pick-up device*

Detail design of SHF dishes depends on the individual manufacturer, but the system is engineered to provide a carrier-to-noise ratio of better than 14 dB for 99 per cent of the worst month weather-wise, with a dish of 60 cm diameter in the primary service area. A larger dish offers improved gain, but is more difficult to align physically.

Offset dishes

All other factors being equal, and so long as the pick-up device is at the focal point, the strength of the received signal is proportional to the *area* of the reflector. For convenience in manufacture, assembly and installation (and perhaps a better aesthetic effect) small fixed-dish installations use an *offset* dish, with the interception electronics mounted on the end of an arm protruding from under the dish as shown in the photograph of Figure 13.6. The principle here is that the dish, no longer parabolic in shape, conforms to the contours of a section of a much larger parabolic reflector, Figure 13.7, with the two advantages that the arm and head unit do not make a 'shadow', signal-wise, on the dish to reduce its efficiency; and that the disc surface is more nearly vertical so that water, snow and other matter is less likely to settle on it.

Figure 13.6 *Offset metal mesh dish for satellite reception*

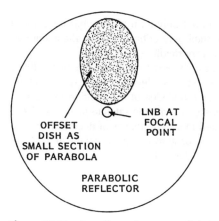

OFFSET
DISH AS
SMALL SECTION
OF PARABOLA

LNB AT
FOCAL
POINT

PARABOLIC
REFLECTOR

Figure 13.7 *An offset microwave dish may be regarded as a small section of a large parabola shape*

Head-end convertors

At the point of pickup, the signal is in 11 GHz FM form, and at that frequency the only possible transmission medium is a waveguide in the form of a large and rigid metal pipe. Any form of co-axial cable would introduce intolerable attenuation, so a frequency converter is mounted at the receiving dipole, in similar fashion to the masthead amplifier of some terrestrial signal-receiving installations. This converts the signal, by means of a heterodyne process, to a lower frequency around 1 GHz which is amenable to feeding indoors by means of very low-loss co-axial cable.

Service areas

Unlike conventional TV transmitters, the coverage of satellite broadcasting is not limited by constraints of line-of-sight, nor (except in a very local sense, as described earlier) by topographical features. For this reason the service area is much more homogenous and predictable for DBS (direct broadcasting by satellite), falling off gradually from the centre of the footprint. Within the primary service area (see Figure 13.1) a 0.6 metre dish should suffice, and in the secondary service area a similar standard of reception can be achieved with a larger (higher-gain) receiving dish, typically 0.9 metre in diameter,. Most parts of the UK are within at least the secondary service area of many European satellite services.

Dish installation

Although dishes for use at 11 GHz are not so large as those required for UK reception of the earlier 4 GHz satellites such as the Russian Gorizont, a fair amount of weight and surface area are inherent in them. This means that their mounting must be very strong and stable, not only from a safety-hazard point of view in strong winds, but also because their very critical angular alignment can only be maintained if they are mounted very rigidly.

Once a stable fix has been achieved the dish must be aligned onto the satellite's beam. The first step is to calculate the azimuth and elevation of the satellite at the receiving site, and this requires an exact knowledge of the position of the site, obtainable from an Ordnance Survey map, or a marine satellite navigator. From this the required angles can be calculated to several places of decimals and the dish aligned with reference to a magnetic or gyro compass. A typical 11 GHz receiving dish and mount are illustrated in Figure 13.6.

Preamplifiers for SHF

We have seen that down-conversion is carried out at the dish by a superhet technique. Before mixing, however, a very low-noise preamplifier must be used to bring the carrier signal up to a suitable level. The noise generated in the first stage virtually determines the S/N ratio for the entire installation, and the performance of this amplifier is thus crucial. The FET (field-effect transistor) is used here, and designs using GaAs FET SHF amplifiers have been developed to offer very low noise figures: less than 0.8 dB at a gain of 17 dB. The tuned circuits for selection and filtering are etched onto a glass-fibre-based PC board.

As SHF front-ends are developed with progressively lower inherent noise, advantage may be taken of this to reduce dish size (making alignment less critical) or alternatively to utilise the same size dish for reception of fringe (outer-footprint) reception of transmissions intended for other areas of Europe. Some readers will be aware that for no perceptible noise on a conventional analogue-derived TV picture an S/N ratio of about 42 dB is required; with FM modulation the carrier-to-noise ratio of 14 dB quoted earlier is quite adequate to provide an acceptable display picture.

The question of interference is a relevant one. Any interference to the up-link is very unlikely, and at the receiving site man-made interference is not significant at SHF, especially with the encoding and modulation systems used. Heavy rain at the receiving site will degrade the signal strength by about 4 dB, but in the primary service area this should not be detrimental to picture quality provided such factors as dish size and alignment and preamplifier gain are adequate.

Satellite TV modulation and encoding

All TV satellites use FM modulation systems. There are three main types of emission: MAC variants, discussed in Chapter 10; digital broadcasting, which we examined in Chapter 11; and FM radiation of subcarrier-based analogue colour signals in Secam, NTSC or PAL, which when demodulated correspond to those of terrestrial transmissions, and require the same sort of post-demodulator processing and colour-decoding circuits as have been described in previous chapters of this book.

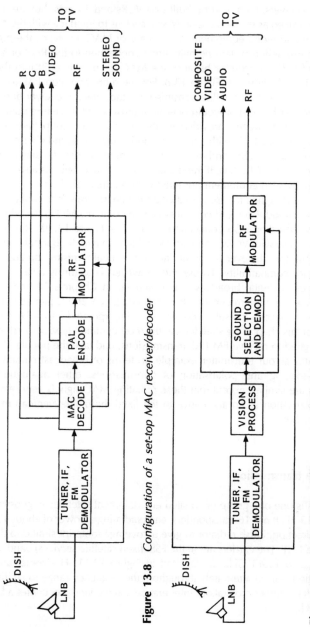

Figure 13.8 *Configuration of a set-top MAC receiver/decoder*

Figure 13.9 *A PAL-FM satellite receiver. The 'vision process' block incorporates a Videocrypt descrambler*

Receiver configuration

Satellite receivers, then, can be built into VCRs and TV sets, but are more commonly produced as stand-alone units with sockets to interface with the AV ports of audio, TV and video equipment. In the case of MAC receivers, the high-quality picture transmission system permits direct connection to the RGB or YUV input sockets of a TV or monitor because the MAC decoder is built into the satellite receiver box as shown in Figure 13.8. For the more common PAL (or Secam or NTSC, for that matter) satellite transmissions, the satellite receiver does not contain a decoder because it would offer no improvement in picture quality, merely duplicating that already provided inside the TV. The set-up here, then, is as depicted in Figure 13.9, with a tuner, demodulator and sound circuits to produce baseband video (CVBS) and audio (stereo where applicable) signals for cable-linking to TV or VCR. A UHF modulator (mono sound only) is also provided for compatibility with equipment not fitted with AV input sockets.

Receivers designed for use with digital satellite transmissions contain, alongside the tuner, demodulator and control system, a complete MPEG decoder and RGB converter, affording RGB connection to the TV – Figure 13.10 – for best picture quality. For use with VCRs, and with TVs which have no RGB input port, they also sport a PAL (or whatever is appropriate to the local terrestrial broadcast system) encoder for compatibility. To use this, however, is to throw away many of the advantages of digital broadcasts, as described in Chapters 10 and 11, and to reintroduce the impairments due to bandsharing of Y and C picture components. The same applies to the subcarrier encoders incorporated into MAC receivers. In the rest of this chapter we shall look at the most common type of satellite receiver, that designed to receive FM PAL transmissions, and take the popular Astra series of satellites as our transmitter example. In terms of broadcast parameters, ERP, channel-spacing and polarisation of transmissions, other analogue-broadcast satellites are similar, except that those radiating MAC signals very often use circular polarisation as an alternative to the fixed (horizontal/vertical) mode to be described below.

Satellite transmission

Astra 1A is one of a group of co-sited entertainment satellites in geosynchronous orbit at 19.2E. It has 16 transponders, each radiating a power of about 45 W into a highly-directional dish radiator to give a coverage footprint similar to that shown in Figure 13.1. The carriers are in the FSS (fixed satellite services) band, and range from 11.2 to 11.45 GHz as depicted in Figure 13.11. Horizontal and vertical polarisation is used alternately throughout the frequency range to facilitate channel overlap without crosstalk or interference. Each channel occupies a bandwidth of 27 MHz.

Figure 13.10 *Internal signal path of a digital satellite receiver*

Figure 13.11 *Carrier spacing and polarisation for a set of satellite transmissions*

LNB

The microwave signals bounced off the dish first encounter a low-noise block (LNB) mounted at the focal point. They are conveyed by a waveguide to a micropatch, whence the vertically- and horizontally-polarised transmissions are coupled to separate printed lands on the PC board, each connected to its own GaAsFET low-noise transistor preamplifier stage, selected as required by the receiver's control system. This selection is typically achieved by switching the LNB supply voltage between 13 V and 17 V: the DC operating voltage for the entire LNB comes up the cable down which the signal passes to the indoor unit.

The operating principle of an LNB is given in Figure 13.12. The RF amplifier uses two or three FETs to provide an overall gain of about 20 dB. After bandpass filtering the SHF signal is applied to a mixer consisting of a Schottky-type diode of either silicon or GaAs construction: its inherent non-linearity provides the mixing/beat action. The local oscillator is again based on a low-noise FET. It works in conjunction with a *dielectric resonator* to produce a very stable and noise-free CW (continuous wave) signal at 9.75 Ghz, on the low-frequency side of the incoming carriers.

Each broadcast vision carrier produces an IF corresponding to the difference between its frequency and that of the local oscillator. Since all the carriers beat simultaneously with the oscillator signal, the result is block-conversion of the whole spectrum of channels to a lower frequency band which is selected by a wideband filter at the mixer's output. The channel spacing and modulation characteristics remain the same. To take as an example Astra 1A ch. 1 at 1 121 425 MHz, the difference frequency produced when it beats against the 9750 MHz local oscillator signal is 1 464 250 MHz or 1.46425 GHz. This passes down the feed-wire as one of the sixteen IF frequencies from Astra 1A. In practice the indoor receiver can tune over the range 950–2050 MHz, which embraces the IFs of all the Astra-group transponders in the several co-sited satellites. The overall gain of the LNB is about 50 dB. A photograph of the internal layout of an LNB appears in Figure 13.13.

190

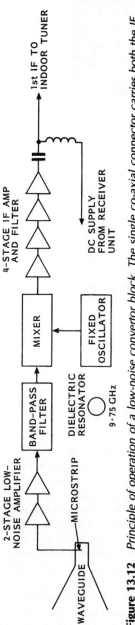

Figure 13.12 *Principle of operation of a low-noise convertor block. The single co-axial connector carries both the IF output signal and the DC operating voltage*

Figure 13.13 *Internal view of an LNB. Micropatch interceptors for H- and V-polarised signals are in the circular land at the top. Many of the print patterns function as bandpass filters*

Indoor tuner and receiver

The indoor tuner selects the satellite TV channels. It's very similar to the UHF tuner described in Chapter 8 – see the block diagram in Figure 13.14. The main difference is that the second IF is much higher in frequency: 140 MHz, 200 MHz or most often 480 MHz. A SAW filter is shown as part of the tuner in Figure 13.14 because in satellite receivers the IF amplifier and FM demodulator are packaged in a single screened module with the tuner section. For this application the SAW filter has a bandwidth of 27 MHz to match that of the transmission. The oscillator within the indoor satellite tuner is governed by a phase-locked loop (PLL) and a reference crystal in a frequency-synthesis tuner system similar to that used in a full-function TV set. The whole operation is orchestrated by a microprocessor, and the tuning data (and many other operating reference points) are stored in a large EEPROM as described in the TV control section of Chapter 8.

Two post-tuner techniques used in satellite receivers remain to be described because they differ from conventional (AM) TV receiver practice. They are FM vision demodulation and energy-dispersal clamping. The FM demodulator is immune to amplitude modulation and impulsive interference so long as its input signal level exceeds a certain threshold. The PLL principle, illustrated in

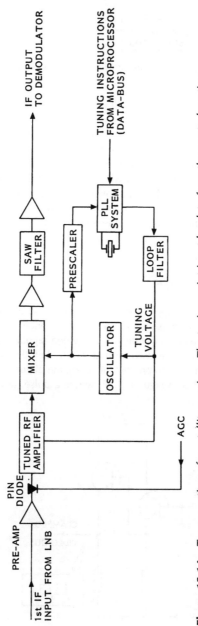

Figure 13.14 *Tuner section of a satellite receiver. The tuning point is set by data from the control section*

Figure 13.15 *Operating principle of PLL FM demodulator*

Figure 13.15, is used in a simple circuit which is invariably buried inside an IC. A linear VCO (voltage-controlled oscillator) free-runs at the second IF, say 480 MHz. Its output forms one feed to a phase detector whose second input is the IF itself, a frequency-modulated carrier. The output from the phase detector, which is an error signal representing the difference between the frequencies of its two inputs, is used to steer the VCO which thus faithfully tracks the vision carrier deviations. The result is that the error voltage, after suitable low-pass filtering, is a facsimile of the original modulating signal: the video waveform complete with the sound carriers and where applicable the teletext data.

Energy-dispersal technique

To prevent fixed patterns appearing in the sideband structure of an FM satellite transmission – and the consequent risk of interference to other services – an energy-dispersal waveform is added to the video signal before modulation. It takes the form of a triangular waveform at 25 Hz, half the field frequency. After demodulation it is removed to prevent flicker on the picture. The action is carried out by a circuit like that in Figure 13.16, where a pulse, timed to coincide with

Figure 13.16 *Video clamp circuit for stabilisation of the black level in a TV signal*

194

some fixed amplitude point in the video waveform (e.g. front porch, black level or sync tip) restores the charge on the right-hand plate of coupling capacitor C1 to a reference potential once per television line.

Encryption

Most satellite picture broadcasts are *encrypted* to give the broadcaster control over who watches them so that subscription systems can operate and so that geographical copyright of programme material is not infringed. The process of encryption, in most cases, consists of changing the timing order in which TV line signals are broadcast, and sending a 'key' along with the TV signal, which carries instructions on how to reassemble them in the correct order to produce the original image. For D-MAC transmissions the encryption system used is called *Eurocrypt*, whose principle is based on individual addressing of receivers (each of which has its own unique identity number, and hence memory-address) over the air.

With the more common PAL/Secam transmission formats used in Europe, different encryption systems are used: the most common of these is *Videocrypt*, which depends on an individually-addressed smart-card sold by the broadcaster and inserted by the viewer into a slot on the front of the receiver. The card contains an IC-based algorithm, in effect a set of instructions held in memory. The Videocrypt concept is to cut each TV line in two, and splice the halves with two other line fragments. There are 256 possible points in each line period where the cut can be made; in a transmission the cut-points are selected in a pseudo-random sequence. The result is a scrambled and totally unrecognisable picture on screen, only sorted out into a coherent picture when the line segments have been reassembled into correct order by a decoder or descrambler. The descrambler is instructed on where the cuts have been made, and the way in which they have been shuffled, by data transmitted (in similar fashion to Teletext) on unused lines in the field blanking interval. The data itself is also encrypted, requiring processing by the algorithm contained in the smart card to define the cut points. The Videocrypt data lines convey 32-byte messages at the rate of about 24 per minute. At the smart card it is converted to a 60-bit value as an instruction to the decoder to generate the cut and shuffle points to reconstruct the picture. Also contained in the 32-byte data are control commands for the individual smart card, defined by its unique address; and a four-byte authorisation code. The former can activate or deactivate the card, and trigger on-screen messages from it, while the latter ensures that only 32-byte instructions from a legitimate source (the broadcaster or his agent) will be actioned by the smart card and descrambler. This elaborate and complex system is designed to prevent unauthorised viewing by 'hackers' and 'pirate' cards.

Satellite sound

The two systems used for broadcast satellite sound are FM and digital. In the case of digital and MAC broadcasts the audio is digitally encoded as described in Chapter 12 and slotted into the digital data stream in similar TDM (time division multiplex) fashion to that illustrated in Figure 10.5. The transmission and reception systems used for PAL/Secam satellite sound were described in Chapter 12.

Receiver architecture

Figure 13.17 gives an idea of what happens inside the satellite set-top unit. The supply voltage for the LNB is switched – to select polarity of reception – by the control system, while its 950–2050 MHz block-converted IF output signal has one of its broadcast components selected by the tuner under the control of the master microprocessor and with reference to the data stored in the EEPROM and the crystal in the frequency-synthesis circuit. These processes are identical to what goes on inside a micro-controlled terrestrial-reception TV set. The second IF output is selected and tuned by a SAW filter, then amplified for application to the FM demodulator.

Emerging now as a baseband signal from the demodulator, the signal takes three paths. It's routed out via a socket for application, if required, to an external decoder, whose descrambled input re-enters for possible selection in a video routing switch operated by the control microprocessor. It also passes, inside the box, to a video de-emphasis network, followed by an energy-dispersal clamp as described above. Now the video signal is processed by the descrambler, to emerge (if the subscription has been paid!) as a 'studio'-state waveform for selection in the routing switch to the video (e.g. 21-pin Scart) output socket and the UHF modulator, which acts as a mini closed-circuit TV transmitter with a carrier output typically on UHF channel 38. The FM sound carriers are filtered out from the demodulated video signal, as described in the previous chapter, and the audio signals they convey are passed to the UHF modulator and to the AV output sockets for connection to the TV/monitor, a VTR, and/or a separate stereo audio system, any of which may incorporate a surround-sound processor for a 'home-cinema' effect.

The microprocessor control system, communication highway, data bus, EEPROM programme/mode memory and (usually switch-mode) power supply system work to exactly the same principles as those described in the appropriate sections of Chapter 8 where their application to terrestrial TV receivers was examined. An internal view of a PAL satellite receiver designed for reception from the Astra group of satellites is shown in Figure 13.18.

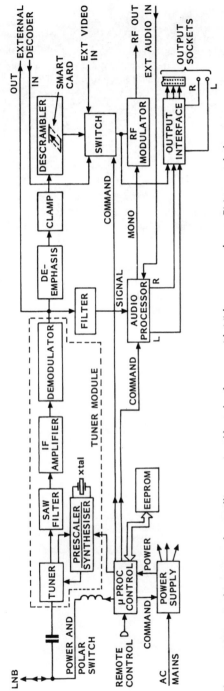

Figure 13.17 *Architecture of a satellite receiver/decoder for use with Videocrypt analogue-FM transmissions*

197

Figure 13.18 *Internal view of an IRD (integrated receiver–decoder) for reception of satellite transmissions. The switch-mode power supply unit is at top right; tuner/demodulator right of top centre; and smart-card reader/descrambler at bottom left (Pace)*

14 *Cable TV*

Originally the philosophy of cable television was simple: to bring television to those who could not receive it by means of conventional RF aerials for reasons of local topography, the impracticability of installing an aerial, or aesthetics as judged by the local council or planning authority. Many CATV (communal-aerial television) systems had their origins in the dark ages of 405-line television on VHF, with few and scattered transmitting sites. With the spread of the UHF terrestrial broadcasting network (transmitters are now being installed for communities of fewer than 300 souls) the necessity for CATV as an alternative to listening to the radio (!) has all but disappeared; blocks of flats, hotels and similar domiciles can be served by MATV (master-aerial TV) which is a small-scale cable system, working from a master aerial and distributing signals to tens, rather than thousands, of TV sets.

In some cases, out-of-area programmes were made available on a community cable network as a 'bonus' to subscribers, although the advent of Channel Four sometimes meant dropping this facility, as many cable systems (especially wired-pair HF and some co-axial VHF networks of long standing) had a maximum capability of four vision channels. Because of government policy, cable operators could, in general, only distribute the programmes of the national broadcasters, and this (in the UK at least) tended to limit the popularity of cable systems. Further problems for old cable networks were the propagation of teletext signals through the network, the difficulty of maintaining good bandwidth and delay characteristics for colour signals and the incompatibility of commercial home VTR machines with the special receivers (called *terminal units*) used with some cable systems.

In the 1980s interest in cable distribution systems was revived with the authorisation by the UK government of twelve pilot schemes for cable TV networks, each to serve about 100 000 households. Hence the great potential foreseen by the early pioneers, but not then permitted to develop, may be realised. The attraction of new cable franchises is that alongside 'off-air' material they are permitted to broadcast exclusive programmes, not obtainable except over the cable, and thus create a demand from the viewing public and a financial incentive for the cable operators. It was envisaged that large-scale 'cabling' of the UK could follow, with its possibilities of Pay TV, and *interactive* services in which the viewer can communicate with a central information and data exchange via the cable, in

199

the same way as Prestel and Internet viewers are able to send and receive video-text over the telephone network. The advent of satellite services opened further prospects for the cable system; those viewers who are unable to accommodate a receiving dish, or who wish to view programmes intended for other European countries (with or without English soundtrack) can be catered for, as the installation of receiving dishes for fringe satellite and DBS (direct broadcasting by satellite) reception is most economically done on a 'community' basis.

Provided the demand (and hence finance) were there, the cable scheme would enable television to become as locally-based as the current BBC and ILR district radio services, particularly relevant in the provision of text and data transmissions of local interest only; and in the potential for local advertising.

Cable types

The early cable transmission system consisting of twisted pairs carrying HF vestigial sideband TV signals is now obsolescent, though used in some districts for satellite programme relay.

The choice for multichannel TV carrying videotext (and possibly other data channels) lies between co-axial, standard telephone and glass-fibre optic cables, and each has various advantages and disadvantages. Co-axial cable is currently well established, and scores on the counts of easy interfacing with existing equipment and lower initial cost, at least on comparatively short runs. The optical-fibre technique requires more complex terminal equipment but has advantages in the areas of data-handling capabilities, immunity to electrical interference, security against 'tapping', and a small physical diameter, enabling a greater number of services to be laid in existing ducts, and a reduction in the cost of routing. Optical fibre has advantages over co-ax in terms of transmission efficiency, too; for a given data-rate the *repeaters* ('boosters' to overcome transmission losses) can be spaced at greater intervals than with the co-ax system. Regarding cost, glass is intrinsically much cheaper than copper and in volume production the glass-fibre technique may well show an overall cost advantage over copper cables.

It has been shown that for trunk lines optical transmission has much to recommend it, and British Telecom currently operates several glass-fibre optic links for transmission of television, audio, data and telephone traffic. Some can operate at very high data rates (140 Mbit/second) which confers the simultaneous ability to handle thousands of phone calls, or many broadcast-quality digital TV channels. It may be that the most economical way to implement a cable system will be to adopt a 'hybrid' solution, with optical-fibre trunk routes to local distribution points, whence co-axial cables will 'spur off' to individual dwellings grouped around the fibre cable head. Alternatively, connection can be made to existing subscribers' telephone lines for conveyance of picture, data and sound.

Transmission modes

In the same way as air or space can be used to carry virtually any radio frequency using any of the several modulation systems outlined in Chapter 1, so it is (within reason!) with co-ax and fibre-optic systems. Obviously the 'launching' and 'interception' methods differ, and for fibre the basic carrier is light, rather than an electrical wave. Thus we can use baseband, AM, FM, PM or PCM in cable systems, at such carrier frequencies as are appropriate to the signal, the distance between terminals, and the transmission medium. Typically a co-axial cable will carry analogue signals modulated by FM or AM onto carriers in the VHF and (for short runs) UHF range. Fibre-optic cables will work from analogue-baseband to the 140 Mbits/second PCM mode described above. It should be remembered that the light-carrier in a fibre system (usually infra-red rather than visible light) is itself an electromagnetic wave with a frequency of the order of 3×10^8 MHz or 300 THz (see Chapter 2) so that the upper limit on the rate of data throughput in an optical fibre, is perhaps, limited by technology rather than physics! The main restriction on bandwidth in glass fibre links using modulation of a sub-carrier (sub-, that is, to the frequency of the light wave itself) is the effect of *fibre-dispersion*, which tends to slightly 'blur' in time the sharpness of received pulses, giving an *integration* effect to their shape.

The network

There are two basic methods of cable distribution. The more traditional is the tree-and-branch system (see Figure 14.1a) in which all available programmes are continuously sent over the network in separate channels, with user selection by means of some form of switch at the receiving point. This was the *modus operandi* of the original radio and TV cable system, in which each household may be regarded as being on the end of a 'twig'.

The alternative and better system is termed a 'switched-star network', shown in Figure 14.1b. Here the available programmes are piped to a 'community-central' point analogous to a telephone exchange. The subscriber can communicate with the 'exchange' and request the desired programme(s) to be switched to his line for viewing, listening or recording. The advantage of the switched-star network is its *interactive* capability, whereby the subscriber can 'talk-back' to a local or central exchange; this opens the possibility of the 'single-fibre' household in which all communications services (radio, TV, telephone, banking, public-utility meter reading, text and data etc) come into the dwelling via a single link which can, by means of recording devices at either end, be utilised during off-peak and night hours. The switched-star configuration lends itself well to Pay-TV (either pay-by-channel or pay-by-programme) because security is more easily arranged. It is simpler to deny a programme to a non-subscriber by a central 'turn-key' process than by expensive signal *scrambling* and 'authorised unscrambling' systems of the sort described in the previous chapter.

Terminal equipment and propagation modes in glass-fibre cables

Since the basic transmission 'vehicle' in fibre-glass cables is light energy we must now see how the signal is launched into the cable and intercepted at the receiving

(a)

(b)

Figure 14.1 *Two methods of signal distribution: (a) the 'passive' trunk-and-branch system; (b) the 'interactive' switched-star network. These are discussed in the text*

end. Depending on the distance to be covered, the sending device may be an LED or low-power semiconductor laser operating on a wavelength (infra-red) around 850 nm. The radiant energy in the sending device is surprisingly small, typically 200–300 µW for an LED and 1–3 mW for a laser. The fibre termination is an integral part of the light-source encapsulation for maximum coupling efficiency, permitting virtually all the light to be concentrated in the cable.

The receiving device is a light-sensitive diode, again intimately coupled to the fibre end. For low noise and highest possible sensitivity this pick-up device is 'tuned' to the light wavelength in much the same way as a radio set is tuned to an RF transmission. For short-haul reception a silicon P-I-N photodiode is generally used; with long-distance fibre cables greater sensitivity can be obtained by using an *avalanche* photodiode which combines the property of light detection with an internal amplification process.

The glass-fibre core is of very small diameter, typically 50–100 µm, surrounded by an intimately-bonded cladding layer of about 20 µm thickness. Further layers give strength and environmental protection, the outer jacket consisting of a tough waterproof polyurethane cover. The transmission of light along a glass fibre depends on the phenomenon of *total internal reflection* in which the light, when it encounters the inner surface of the fibre wall, is 'bounced' by the mirror-like wall surface back into the fibre. Light can enter the fibre end at any angle and the bounce-path in transmission can thus take several forms, as shown in Figure 14.2. Large bounce-angles give rise to a long path length (in terms of the passage along the entire cable) and are called high-order transmission modes; at lesser angles the total path length is shorter, described as low-order transmission mode. A beam which travels down the axis of the fibre takes the shortest possible path length in what is known as axial mode.

The nature of light propagation down a fibre depends on its diameter and on the difference in refractive index between the fibre core and its cladding material. Where a sudden change of refractive index is present at the fibre wall we have a *step-index* fibre in which several modes (high- and low-order) are taken by the light. An alternative form of construction is the *graded-index fibre*, where the interface between 'core 'and 'cladding' represents a more gradual change in refractive index. This has the effect of making the light rays turn less sharply when they encounter the fibre wall and thus reduces reflection loss, as shown at the top right of Figure 14.2. Low-order modes predominate in such a graded-index fibre cable, and such high-order modes as are present travel faster along their longer path, reducing the fibre-dispersion effect described earlier. Graded-index fibres offer a low transmission loss and greater bandwidth than step-index types, at the expense of higher production costs and greater coupling losses at the junctions between the fibre and the sending and receiving devices.

Both step- and graded-index fibres operate in what may be called multimode with many possible light path angles within the fibre. If we can arrange a fibre to concentrate on the axial mode we shall significantly reduce the transmission loss. In this *monomode* system a high-grade glass core is used, with a small diameter in the region of 5 µm (about one-tenth of diameter of multimode cable cores). The light wavelength used here is longer, around 1.35 µm, and the much straighter light path gives very good transmission efficiency. Repeaters (regenerators) are therefore required at much longer intervals than in conventional fibre (and particularly co-axial) systems, and in a typical monomode transmission system repeaters can be as much as 30 km apart; this economy in equipment easily outweights the cost disadvantage of the monomode fibre cable itself.

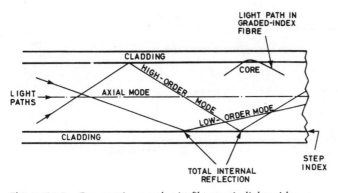

Figure 14.2 *Propagation modes in fibre-optic lightguides*

Repeater power

For repeaters generally, the operating power can be sent down the cable in co-axial systems, as described earlier for domestic masthead amplifiers. Fibre-optic cable plainly cannot carry DC, but conductive members can if required be incorporated in its protective sheath, or local power sources can be used, in view of the long intervals between repeaters made possible by fibre-optic technology. Many thousands of kilometres of fibre-optic cable are in use in the UK, primarily by British Telecom.

TV via the phone line

In the UK alone there are over 30 million twisted-pair phone lines in operation between British Telecom exchanges and individual subscribers' premises. Most of them use copper wire, some aluminium, and all represent a ready-made medium for distribution of TV signals if they can be pressed into service in this application.

Originally the telephone lines were designed to carry simple command (dialling) pulses and frequency-restricted (300–3500 Hz) baseband voice signals, and their attenuation factor at higher frequencies is very great: that of a typical voice-grade telephone circuit at 200 kHz is 60 dB, and at 1 MHz is 110 dB. The *group delay* characteristic, however, is constant over a wide range; group delay is an expression of the ratio of phase change per unit of frequency. Under these circumstances pulse signals suffer relatively little distortion, opening the way to transmission of relatively high bit-rate digital datastreams over conventional 'phone cables.

With the type of digital compression (MPEG-standard) described in Chapter 11, a bit-rate of 2 Mbit/second can be used to send an entertainment-standard (equal-to-VHS) down a telephone line, complete with high-quality digital stereo sound. At a bit-rate of 4 Mbit/second broadcast-standard pictures are possible, and in both cases provision is made for a relatively low bit-rate return channel via which the subscriber can specify and request the service, movie etc. required, and even to provide such features as freeze-frame, search forward or backward (cue/review) and slow motion. The spectrum allocations for this are given in Figure 14.3. There

Figure 14.3 *Spectrum of digital TV signal in a narrowband telephone circuit*

Figure 14.4 *DMT modulation system for narrowband transmission channel. Adaptive separation techniques can be used to optimise signal transfer*

are several ways of configuring the control and picture/sound data, depending on the length and electrical characteristic of the line in use. Where required, it can be done *adaptively* by automatically selecting the best mode for optimum S/N ratio and minimum bit error rate. The data passes over the line in balanced form to minimise radiation and RF interference, coded in a format called 2B1Q wherein bit-pairs are individually converted to one of four quaternity symbols.

Modulation

Various modulation systems can be used to convey the data; an effective one is DMT (discreet multi-tone modulation) in which the data is distributed among many sub-channels in similar fashion to the OFDM (orthogonal frequency division multiplex) scheme described in Chapter 11. Figure 14.4 outlines the process: the serial input datastreams are first encoded in parallel form, then the resulting frequency samples are converted into time values in an FFT (fast Fourier transform) processor. Now the derived values are transcoded into serial form for application to a D/A converter before transmission. Noise and interference picked up by the signal in its passage is distributed over many channels by this technique, and the resulting minimal damage to individual pulse trains is amenable to repair by the sorts of techniques described earlier in this book in connection with digital sound and vision processing. Error control by cross-interleaving is incorporated into the transmission code to assist the process.

15 *Development of video tape recording*

The story of video tape recorders really began before the turn of the century with the experiments of Valdemar Poulson. By this time the relationship between electricity and magnetism was well understood, and the idea of impressing magnetic pulses on a moving magnetic medium was sufficiently advanced in 1900 to justify the US patent on Poulson's apparatus, the Telegraphone. The medium was magnetic wire rather than tape, and without any form of recording bias, and scant means of signal amplification, the reproduced sound signal was low, noisy, non-linear and lacking in frequency response. These problems of 'tape' and record/replay head performance are ones that have continually recurred throughout the history of sound and vision tape recording, as we shall see.

By the early 1930s, many advances had been made in the field. DC bias or pre-magnetism of the recording wire had been tried with better results, then overtaken by the superior system of AC bias, as used today. The magnetic wire gave way to steel tape 6 mm wide travelling at 1.5 metres per second, and performance became comparable with the contemporary disc recording system. Not 'hi-fi' by any means, but certainly adequate! The BBC adopted and improved the Blattnerphone system and in 1932 broadcast a programme of the Economic Conference in Ottawa, for which *seven miles* of steel tape was used, edited by means of a hacksaw and soldering iron! This era also saw the first crude forerunner of a servo system in the Marconi–Stille machine of 1934. The earthed tape was arranged to contact insulated metal plates when it became slack, the plates being wired to thyratron control valves. Relays in the thyratron anode circuits modified drive motor currents to regulate tape speed.

A great impetus was given to the industry when it became possible to coat a flexible insulated base with a finely-divided magnetic substance. This was achieved in Germany by Dr. Pfleumer and developed by Wilhelm Gaus under the auspices of AEG. This activity culminated in the successful demonstration at the 1935 Berlin Radio Exhibition of the first commercial sound tape recorder, the AEG Magnetophone, using a cellulose acetate tape coated with carbonyl iron powder. Performance of these sound recording machines steadily improved during the '30s and '40s to the point where at the end of the 1940s much radio broadcast material was off-tape, and indistinguishable from live programmes.

Once audio magnetic recording had become established in the radio industry, attention was turned towards the possibility of recording television images on tape. The problems were formidable, mainly because of the relatively large bandwidth of a television signal. Plainly, it would be necessary to increase the tape speed and one approach, by Crosby Enterprises in 1951, took the form of a 250 cm/second machine designed to record a monochrome picture whose frequency spectrum was split into ten separately-recorded bands with additional sync and control tracks. The same principle was embodied in a fearsome machine designed by RCA, in which the tape travelled at 600 cm/second to record on three separate tracks simultaneous R, G and B information for colour TV. Other designs involving longitudinal recording called for tape speeds approaching 1000 cm/second, amongst which was the BBC's VERA (video electronic recording apparatus) of 1956. Such machines were wasteful of tape and thoroughly frightening to anyone who happened to be in the room in which they were operating!

Already the seeds had been sewn of a new system, one which was to hold the key to the modern system of TV tape recording. This was simply the idea of moving the record or replay heads rapidly over the surface of a slowly-moving tape to achieve the necessary high 'writing' speed. Initially, the hardware consisted of a circular plate with three recording heads mounted near its edge at 120° intervals, their tips protruding from the flat surface of the faceplate. The 5 cm-wide tape was passed at 76 cm/second over the surface of the rotating plate, whose heads had an effective velocity of over 6000 cm/second, resulting in narrow arcuate tracks across the width of the tape. It was a step in the right direction, but results were poor for several reasons. The bandwidth of the TV signal being recorded was difficult to get on and off the tape due to noise and head-gap problems (we will meet these in detail in the next chapter), and the valve technology of the time did not really lend itself to such requirements as an ultra-wideband, high gain and stable amplifier.

Two more factors were required for success, and these were engineered by Dolby, Ginsburg and Anderson, of Ampex between 1952 and 1955. The problems associated with the tape-track configuration were solved by the use of a horizontal head-drum, containing four heads and rotating on a shaft mounted parallel to the direction of tape motion as shown in Figure 15.1a. The video heads lay down parallel tracks across the tape width, slightly slanted, (see Figure 15.1b) with each head writing about 16 television lines per pass. The tape-to-head speed (i.e. the *writing speed*) at an incredible – to us today – 40 metres per second was most adequate, with response beyond 15 MHz. The final hurdle was cleared with the introduction of an FM recording system; this involves frequency-modulating a constant-amplitude carrier with the picture signal before application to the recording head. Charles Ginsberg, one of the collaborators, once said that FM was proposed by the man who was assigned to design an AGC circuit for the AM system originally in use! The Ampex *Quadruplex* system was enthusiastically taken up by TV broadcasters, and rapidly became a world standard. For less exacting requirements and limited funds, a simpler recording system was really needed, and this gave birth to the helical-scan system. The idea of a rotating video-head drum containing one or two heads, each laying down one complete television field per pass, was first mooted in 1953. The Japanese Toshiba company were foremost in the field in the early days, though several

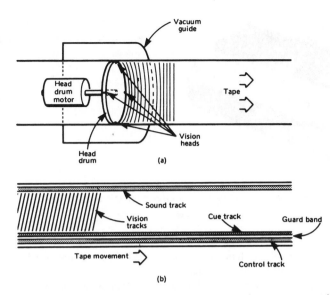

Figure 15.1 *(a) The transverse-scan system. The vacuum guide draws the tape into a curved profile to match the contour of the head drum. (b) Resulting tracks on the tape*

other companies were working on the idea. By 1961 a handful of manufacturers were demonstrating helical machines, all with the open-reel system, each totally incompatible with all the others, and none having a performance which could approach that of the quadruplex system. The market for these machines was intended to be in the industrial and educational spheres, with very little regard as yet to the domestic market. If there was one thing these professional users wanted above all else, it was standardisation of the helical format to ensure software compatibility. Does this, 35 years later, have a familiar ring about it?

By the early 1970s, helical-scan machines had, by and large, settled into three camps. The Sony U-matic standard was well established, with a very good performance for its era, using 12.7 mm tape in a large cassette for easy handling. Most Far-Eastern manufacturers had thrown their weight behind the EIAJ (Electronics Industries Association of Japan) system, using a cartridge containing a single spool of tape. The third contender was Philips. Their 'VCR' machine, illustrated in Figure 15.2, was the first to be designed truly for the domestic as well as the educational and industrial markets. It took a relatively compact, fully-enclosed cassette and was styled for the home environment. Easy-to-operate controls, a simple timer facility and one-hour capability were offered, together with built-in TV tuner and RF modulator for use with a conventional TV set. This, the N1500, was a milestone in VTR development.

The N1500 did not sell in very great numbers, in spite of the intensive effort made by Philips' engineering and publicity departments. Its release in 1972 coincided with an explosion in colour TV sales, and the average householder blew his savings on a colour receiver rather than a VTR. The maximum one-hour playing

Figure 15.2 *The Philips VCR format was first in the domestic field. A late version of the first Philips machine, type N1500*

time, insufficient for a football match or a feature film, also held back sales and the introduction by Philips in 1976 of a similar machine, but with two-hour capability (N1700, VCR-LP) was a step in the right direction. The Philips monopoly was broken, however, in 1978 when the rival Japanese systems VHS (Matsushita) and Betamax (Sony) appeared on the shelves of high street shops.

All three systems used cassettes, a thick double-stacked type for the Philips VCR and VCR-LP format, and smaller, lighter, co-planar cassettes for the VHS and Betamax systems. All the early machines of each format had piano-key controls similar to those of audio tape recorders. The rather awkward and expensive Philips VCR cassette, together with its relatively short playing time, spelt the early demise of this format, and by 1980 VHS ruled supreme in the UK, with Betamax in second place and rapidly gaining ground. Sales of domestic VTRs rapidly increased, and a wide range of software appeared alongside. The pace of machine development quickened, and in 1980 piano-keys had given way to light-touch sensors and remote control, sophisticated timers and programmers had appeared, and a form of freeze frame had been introduced. Another significant event of 1980 was the appearance of a new helical VTR format, the Philips/Grundig Video 2000 system, for use with the VCC (video compact cassette). The Compact Cassette designation indicated the intention that it should become as popular as the universally-used audio compact cassette. In the event V2000 format was not successful, and production had ceased by early 1986. It used an advanced DTF (dynamic track following) tracking system, very similar to that of the later and more successful Video 8 format, whose *ATF* feature will be described later.

Specialised ICs had been making steady inroads into domestic VTR design, LSI (large scale integration) and microprocessor devices becoming commonplace in contemporary machines. These made possible advanced remote control systems, comprehensive timers and programmers, trick-speed, still-frame and visual search features. They also led a trend away from mechanical complexity in the tape deck and towards electronic control of direct-drive systems; this considerably simplified the mechanics of the tape transport, threading and head drive systems, while retaining a relatively low electronic component count.

Advances have been made, too, in the field of portable VTR equipment. Purpose-designed battery operated machines became available in VHS, VHS-C and Video-8 format for outdoor location work. Those currently on offer incorporate the camera section and videorecorder in one unit – *camcorders*. In this realm there is strong competition between the VHS camp whose champions are JVC and Panasonic, and the Video 8 protagonists, led by Sony of Japan and having in its ranks many companies with backgrounds in the world of conventional and cinephotography. Details of VHS, VHS-C and Video-8 formats will unfold throughout the book, and the relative merits of the formats (including the high-definition variants, S-VHS and Hi-8) will be discussed in Chapter 26.

16 *Magnetic tape basics and video signals*

All magnetic materials, videotape coatings amongst them, may be regarded for practical purposes as consisting of an infinite number of tiny bar magnets, each with its own north and south poles. This is a simplification, but suits our purposes well. In the natural state these bar magnets are randomly aligned within the material so that their fields cancel one another out, and no external magnetic force is present. Thus the contents of a box of steel nails, for instance, will have no particular attraction for each other. To magnetise the material, be it a blank tape or a solenoid core, we have to apply an external magnetic force to align the internal magnets so that they sit parallel to one another, with all their N poles pointing in the same direction. When the external field is removed, most of the magnets remain in alignment and the material now exhibits magnetic properties of its own.

Remanent magnetism

If the relationship between externally applied force, or *flux* and retained flux in the material, were linear, the business of tape recording would be much simpler. Unfortunately this relationship, called the *transfer curve*, is very far from linear, as Figure 16.1 shows. Here we have a graph with the magnetising force (H) plotted along the horizontal axis, and the stored flux density (B) on the vertical axis. Our starting point (with the new box of nails!) is in the centre, point 0. Here, no external magnetising force is present, and the magnetic material's internal magnets are lying in random fashion – hence zero stored flux. Let's suppose we now apply a linearly-increasing external flux. The *flux-density* (strength of magnetism) in the material would increase in non-linear fashion as shown by the curve 0-T. At point T the material has reached magnetic saturation and all its internal magnets are rigidly aligned with each other – no increase in applied magnetic force will have any effect. If we now remove the magnetising force, bringing the H coordinate back to zero, we see that a lot of magnetism is retained by the specimen, represented by point U. This is the *remanence* of the material, the 'stored charge' as it were. For recording tape, it needs to be as high as possible.

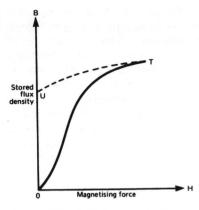

Figure 16.1 *Initial magnetisation curve of a ferro-magnetic material*

Hysteresis

Figure 16.2 is an expansion of Figure 16.1 to take in all four quadrants. We left the material at point U with a stored flux. To remove the flux and demagnetise the material (we'll call it tape from now on) it's necessary to apply a negative magnetising force, represented by 0–V, whereupon stored flux B returns to zero.

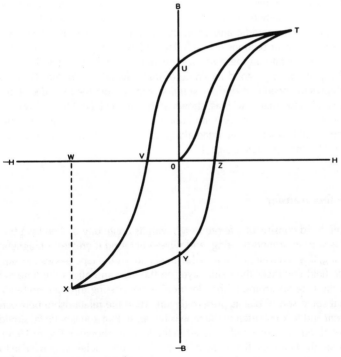

Figure 16.2 *Hysteresis loop*

213

Further negative applied force (V–W) pushes the stored flux to saturation point in the opposite direction, so that all the magnets in the tape are once again aligned, but all pointing the other way, represented by point X. Removal of the applied force takes us back to the remanence point, this time in the negative direction – point Y. To demagnetise the tape a positive applied force 0–Z is required, an increase of which will again reverse the stored flux to reach saturation once more at point T. The diagram is called a magnetising hysteresis curve, and one cycle of an applied magnetising waveform takes us right round it. It can now be seen how the 'degaussing' or *erasing* process works. Here we apply a large alternating magnetic field, sufficient to drive the material to saturation in both directions. The field is then allowed to decay linearly to zero, creating smaller and smaller hysteresis loops until they disappear into a dot at point 0, and the material is fully demagnetised.

Transfer characteristic and bias

Figure 16.3a is based on the previous diagram, but shows the initial magnetising curve in solid line (T0X) with the hysteresis curve in dotted outline. If we apply a magnetising force, 1H, and then remove it, the remanent flux falls to a value 1'. A larger magnetising force 3H will, if applied and removed, leave a remanent force 3', and so on, up to saturation point 8H and remanence point 8'. The same applies in the negative direction, and plotting remanence points against applied force we get the curve shown in Figure 16.3b. This is the transfer characteristic. It is very non-linear at the middle and ends, but for a typical recording type, will contain reasonably linear sections on each flank, PP and QQ. If we can bias the recording head to operate on these linear parts of the transfer characteristic, the reproduced signal will be a good facsimile of that originally recorded. DC bias (Figure 16.4) puts the head into *one* of the linear sections, but the recorded signal will be noisy and inadequate; AC bias, shown at the top of Figure 16.4, allows the head to operate in two quadrants, with superior results. We shall see that the chrominance signal in a VTR is recorded along with a bias signal, which is in fact the FM carrier for the luminance signal.

Head-tape flux transfer

The recording head consists of a ferrite 'ring', with its continuity broken by a tiny gap. A coil is wound around the ring, and when energised it creates a magnetic field in the ring; this is developed across the head gap. As the tape passes the gap, the magnetic field embraces the oxide layer on the tape and aligns the 'internal magnets' in the tape according to the electrical signal passing through the head. Provided that some sort of bias is present (Figure 16.4) the relationship between *writing current* and flux imparted to the tape is linear, so that a magnetic facsimile of the electrical signal in the head is stored in the tape, as shown in Figure 16.5a. The tape passes the head at a fixed speed, so that low frequencies will give rise to long 'magnets' in the tape, and high frequencies short ones.

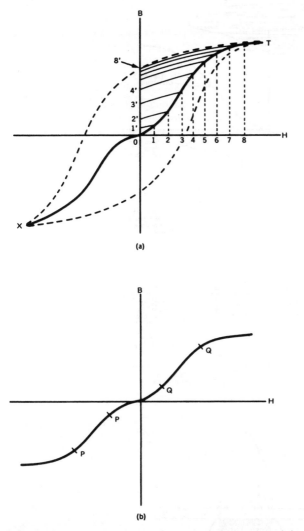

Figure 16.3 *Deriving the transfer curve: at (a) is plotted the remanent flux for eight linear steps of applied magnetising force; (b) shows the resulting transfer curve*

Head gap and writing speed

The linear relationship between head field strength and stored flux in the tape, described above, holds true when the wavelength to be recorded on tape is long compared to the width of the head gap. However, when the wavelength of the signal on the tape becomes comparable with the head-gap width, the flux imparted to the tape diminishes, reaching zero when the recorded wavelength (or rather the non-recorded wavelength!) is equal to the width of the head gap.

Figure 16.4 *The effect of AC and DC bias*

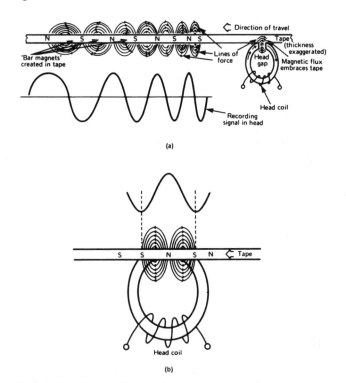

(a)

(b)

Figure 16.5 *Storing flux on the tape: at (a) the flux appearing across the head gap is penetrating the oxide surface to leave magnetic patterns stored on the tape; (b) shows the effect when one complete cycle of the recorded waveform occupies the head gap – no signal transfer will take place. This point represents fex, extinction frequency*

This is illustrated in Figure 16.5b, where it can be seen that during the passage of a single point on the tape across the head gap, the applied flux has passed through one complete cycle, resulting in cancellation of the stored flux in the tape.

This point is known as the extinction frequency ($f$ex) and sets an upper limit to the usable frequency spectrum.

For video recording we need a large bandwidth, and a high $f$ex. This can be achieved by reducing the head-gap width or alternatively increasing head-to-tape, or *writing* speed. There is a practical limit to how small a gap can be engineered into a tape head, and currently this is about 0.3 μm, less than half a micron (1 micron = 10^{-6} metre). For domestic applications, where full broadcast-bandwidth signals are not required, a writing speed of about 5 m/second is required with such a head gap. How this is achieved will be explained in due course. Other HF losses also occur during recording. The head is by definition inductive, so losses will increase with frequency. Eddy currents in the head will add to these losses, as will any shortcomings in tape-to-head contact. High frequencies give rise to very short 'magnets' in the tape itself, and it is the nature of these to tend to demagnetise themselves. For all these reasons, the flux imparted to the tape tends to fall off at higher frequencies, as in the solid line of the graph shown as Figure 16.6. To counteract this, recording equalisation is applied by boosting the HF part of the signal spectrum in the recording amplifier, as per the dotted line.

This is called recording equalisation, and its aim is to make the frequency/amplitude characteristic of the signal stored on the tape as flat as possible.

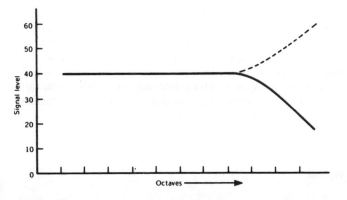

Figure 16.6 *Losses in the recording process. The dotted line shows a compensating 'recording equalisation' curve*

Replay considerations

As with any magnetic transfer system, the output from the replay head is proportional to the 'rate of change' of magnetic flux. Thus, assuming a constant flux density on the recorded tape, the replay head output will double for each doubling of frequency. A doubling of frequency is called an *octave* and a doubling of

Figure 16.7 *Playback curve for a tape with equal stored flux at all frequencies. For a 'level' output the gain of the replay amplifier must follow the dotted replay equalisation curve*

voltage represents an increase of 6 dB. This holds good until the extinction frequency, *f*ex, is approached, when the head output rapidly falls towards zero. This is shown in Figure 16.7. The upper limit of the curve is governed by the level of the signal recorded on the tape, limited in turn by the tape's magnetic saturation point. At the low-frequency end, the replay head output is low due to the low rate of change of the off-tape flux. At some point, it will be lost in the 'noise' off-tape, and this will occur at about 60 dB down from peak level. Thus, even with play-back equalisation (represented by the dotted curve in Figure 16.7), the *dynamic range* of the system is confined to 60 dB or so. With the unalterable 6 dB/octave characteristic, we are limited, then, to a total recording range of ten octaves. This is inherent in the tape system and applies equally to audio and video signals. Ten octaves will afford an audio response from 20 Hz to 20 kHz, which is quite adequate. TV pictures, however, even substandard ones for domestic entertainment, demand an octave range approaching 18 and this is plainly not possible. No wonder they had so much trouble in the pioneering days!

Modulation system

To be able to record a video signal embracing 18 octaves or more it is necessary to modulate the signal on to a carrier and ensure that the octave range of the carrier is within the capabilities of the tape recording system. While the carrier could be AM (amplitude modulation), the FM (frequency modulation) system has been adopted because it confers other advantages, particularly in the realm of noise performance. An FM signal can be recorded at constant level regardless of the modulating signal amplitude, so that head losses and effects of imperfect head-to-tape contact are less troublesome. To achieve a picture replay with no perceptible background noise (snow), the S/N ratio needs to be about 40 dB, and this can be achieved by an FM recording system in a domestic VTR. Professional and broadcast machines can do much better than this!

218

FM basics

An FM system, familiar to us in VHF sound broadcasts, starts with a CW (continuous wave) oscillator to generate the basic carrier frequency. The frequency of the oscillator is made to vary in sympathy with the modulating signal, audio for VHF sound transmitters, video for VTR recording systems and satellite broadcasts. For any FM system the *deviation* (the distance that the carrier frequency can be 'pulled' by the modulating signal) is specified. In VHF sound broadcasting it is ±75 kHz, giving a total frequency swing of the carrier of 150 kHz. In a VTR, carrier frequencies are specified for zero video signal amplitude – represented by the bottom of the sync pulse – and full video signal amplitude, i.e. peak white.

All modulation systems generate sidebands, and those for FM are more complex than occur in an AM system. In fact an FM system theoretically generates an infinite number of sidebands, each becoming less significant with increasing distance from the carrier frequency. Figure 16.8 shows the sidebands of a VHF-FM sound broadcast transmission. The modulating frequency is 15 kHz, and the sideband distribution is such that the first eight sidebands on either side of the carrier

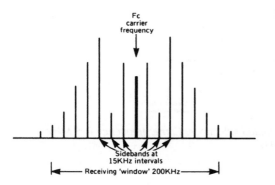

Figure 16.8   *Sidebands of a VHF-FM sound transmitter*

are significant in conveying the modulation information. Thus to adequately receive this double-sideband transmission we need a receiving bandwidth, a window, of 240 kHz or so – in practice 200 kHz is sufficient, and this is the allocated channel width.

Modulation index

The example given, 200 kHz bandwidth for transmission of a 15 kHz note, seems very wasteful of spectrum space, and certainly will not do for our tape system in which elbow-room is very limited! In the above example, the *modulation index*, given by the formula

$$\frac{\text{carrier deviation}}{\text{modulating frequency}} \quad \text{is} \quad \frac{75}{15} \quad \text{or} \quad 5.$$

If we can reduce the modulation index, the significant sidebands draw closer to the carrier frequency, and at modulation indexes below 0.5, the energy in the first sideband above and below the carrier becomes great enough for them to convey all the necessary information in the same way as those of an AM signal. In video tape recording we go a step further and use only one sideband along with a part of the other, rather similar to the vestigial sideband scheme used with AM television broadcast systems.

By using a low modulation index, then, the sidebands of the FM signal can be accommodated on the tape. FM deviation has to be closely controlled and carrier frequencies carefully chosen to avoid trouble with the sidebands, which if they extend downwards from the carrier to a point beyond zero frequency will not disappear, but 'fold back' into the usable spectrum to interfere with their legitimate fellows, leading to beat effects and resultant picture interference. The effect of a folded sideband is shown in Figure 16.9.

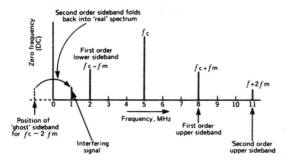

Figure 16.9 *The effort of folded sidebands. Carrier frequency is 5 MHz, modulating frequency 3 MHz*

FM video

Taking the VHS system as an example (all the formats use similar frequencies and parameters), the video signal is modulated on to the FM carrier according to Figure 16.10.

Sync-tip level gives rise to a frequency of 3.8 MHz, black-level 4.1 MHz, mid-grey 4.45 MHz and peak white 4.8 MHz. Thus deviation is limited to a total of 1 MHz, and with a restricted video bandwidth of about 3 MHz, modulation index is about 0.3. The resultant spectrum of the on-tape signal is shown in Figure 16.11, with a full lower-sideband, vestigal upper sideband (limited by the system frequency response and the approach of $f$ex) and a carefully arranged gap between 0 and 1 MHz into which (as later chapters will show) is shoe-horned the chrominance, and where applicable ATF (automatic track-finding) control signals. If we regard the FM luminance spectrum as extending from 1 MHz to 6 MHz, it is now occupying less than three octaves, well within the capabilities of the system. The S-VHS parameters will be discussed later.

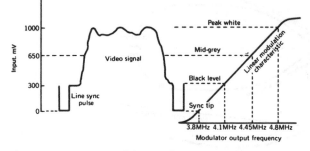

Figure 16.10 *FM modulation characteristic for a VTR*

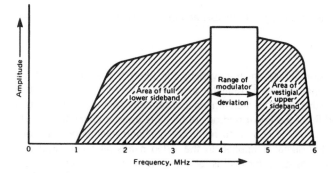

Figure 16.11 *Spectrum of luminance video signal on tape*

Pre-emphasis

Noise is the enemy of all recording and communications systems, and in FM practice it is common to boost the HF components of the modulating signal prior to the modulation process. FM radio uses this technique, called pre-emphasis, and so do we in VTR FM modulation circuits. The effect of this after demodulation (in the post-detector circuit of a radio or the playback amplifier of a VTR) is to give a degree of HF lift to the baseband signal. In removing this with a filter, gain is effectively reduced at the HF end of the spectrum, with an accompanying useful reduction in noise level. The idea is shown in Figure 16.12, which illustrates the filter charactertistics in record and replay, along with the effect on the noise level.

Summary

The FM system is fundamental to the video recording process. Here is a summary of the main points, as applied to domestic VTRs. The video signal is bandwidth-limited in a low-pass filter to restrict it to the band 0–3 MHz. This luminance

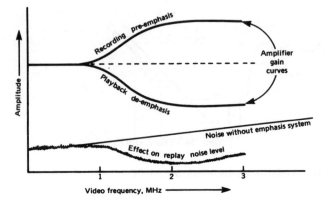

Figure 16.12 *The effect of pre-emphasis and de-emphasis on playback signal-to-noise ratio*

signal (chroma components are removed by the filter) is pre-emphasised and applied to a voltage-controlled oscillator to provide an FM signal with a total swing of about 1 MHz, which is then limited and carefully controlled in amplitude to fall in the middle of the linear portions of the tape transfer characteristic. The FM oscillator frequency is chosen to minimise sideband interference and beat effects, and the FM signal is applied to the video heads after equalisation to compensate for HF losses. The signal is recorded on the tape at a writing speed of around 5 m/second, and the recorded tape carries a vestigial sideband signal ranging up to 6 MHz or so.

17 Video tape: tracks and transport

In Chapter 15 we saw that the necessary high head-scanning speed was first achieved by fast longitudinal recording techniques, then in the successful Quadruplex system by transverse scanning of the tape by a four-head drum with the video heads mounted at 90° intervals around its periphery. The transverse-scan method has a lot going for it! The very high writing speed confers great bandwidth, enabling full broadcast-specification pictures to be recorded and replayed. The relatively wide tracks are almost at right-angles to the tape direction, so that any jitter or flutter in the tape transport has little effect on the timing of the video signal, merely causing slight momentary tracking inaccuracies, easily catered for by the wide track and the FM modulation system in use.

The cost of transverse-scanning machines is high because of the need for a complex head-drum system and precision vacuum tape guides. To maintain the necessary intimate contact between video head and tape, all VTRs have their head tips protruding from the surface of the drum so that they penetrate the tape and create a local spot of 'stretch.' Hence the need for a precise vacuum guide at the writing/reading point in a transverse machine to maintain correct tape tension. These techniques are not amenable to domestic conditions or budgets, and the alternative and simpler helical system has undergone much development. Helical performance has reached a stage where it has challenged the transverse system in professional and broadcast fields, and the helical principle is embodied in all home VTR formats.

Principle of helical scan

In a helical scan system, several problems are solved in one go, but other shortcomings are introduced. The idea is to wrap the tape around a spinning head drum, with entry and exit guides arranged so that the tape path around the head takes a form of a whole or part-helix. The principle is shown in Figure 17.1 where a video head in the course of one revolution enters the tape on its lower edge, lays down a video track at a slant angle and leaves the upper edge of the tape ready to start again at the tape's lower edge on the next revolution. During the writing of

223

Figure 17.1 (a) The Omega wrap. This diagram shows the arrangement for a single-head machine; the path of the head is the heavy line in (b)

Figure 17.2 The diagram of Figure 17.1 drawn from the point of view of the tape

this one track, the transport mechanism will have pulled the tape through the machine by one 'slant' video-track width, so that track no. 2 is laid alongside the first; each track is the width of the video head. Figure 17.2 shows how this works and gives an elementary impression of the track formation.

Let's look at the strengths and weaknesses of the helical format. Tape tension and hence head-tip penetration is now governed by the tape transport system rather than a precision guide arrangement. One or two complete television fields can be laid down per revolution of the head drum, so that the problems of matching and equalising heads (to prevent picture segmentation) disappear. In a single-head machine such as we have described, no head switching during the active picture period is necessary. Two-head helical machines have a simple head-switch or none at all; what price do we pay for these advantages? The main penalty is timing jitter in the recorded and replayed video signals. Because the tracks are laid at a small angle to the tape direction (about 5° off horizontal for VHS) they may be regarded as virtually longitudinal, so that the effects of the inevitable transport flutter, variations in tape tension, bearing rumble etc. will be to introduce minute timing fluctuations into the replayed signal. This effect cannot be eliminated in any machine, and causes problems with colour recording and certain types of TV receiver, as will become clear in Chapters 19 and 25.

In practice, two video heads are used in the drum of a helical VTR and they can be seen in the photograph of Panasonic (VHS) head drum (Figure 17.3) The two-head system means that the tape needs only to be wrapped around half the video head drum perimeter, with one head joining the tape and beginning its scan as the other leaves the tape after completing its stint. To give a degree of overlap

Figure 17.3 *Video head drum. This photograph of the underside of a video head drum shows the head chips protruding from the drum surfaces. This is a four-head drum for VHS format by Panasonic. Note the two windings (and two gaps) per chip*

Figure 17.4 *Omega wrap for a two-head drum. The tape occupies rather more than half a turn of the drum*

between the duty-cycles of the two heads, the tape wrap is in fact about 186°, slightly more than half a turn. This is known as an omega (Ω) wrap, outlined in Figure 17.4.

Track configuration

A typical track layout for a two-head helical VTR appears in Figure 17.5. Here we can see the video tracks slanting across the tape, shaded for head A, white for head B. At the edges of the tape, further tracks are present: the lower carrying a control track (serving a similar purpose to the sprockets in a cine film, and described later) and the upper carrying the sound track. Sound is recorded longitudinally in the same way as in an audio recorder, but with limited frequency

Figure 17.5 *Video tracks written by a two-head system, with each head writing alternate tracks. The relative positions of the sound and control tracks are also shown (VHS format)*

response due to the low linear tape speed in the sorts of VTR we are dealing with. Budget machines only have this mono longitudinal sound track; other VTRs are additionally fitted for hi-fi sound. This superior system is described in Chapter 20. Tape is 12.65 mm (VHS) or 8 mm (Video 8) wide. It progresses at speeds between 23.4 and 10.06 mm/second, depending on format and mode.

Now let us examine the practicalities of the mechanical arrangements of helical scan machines.

Tape threading

One characteristic of all domestic VTRs is the storage and transport of the tape in an enclosed cassette. This not only prevents contamination of the tape and physical damage to it, but also makes easy the loading and operation of the machine. Unlike an audio cassette, where the capstan, pinch-wheel (pressure roller) and heads 'come to visit', as it were, the video cassette system requires that a loop of tape be drawn from the cassette and wrapped around the video head drum, stationary heads, tape guides and capstan assembly. This is called threading, and of the several ways of going about it, we shall describe two.

Domestic VTRs use, as we have seen, a co-planar cassette similar in form to the compact audio type, but larger. To load the machine, the cassette is fed horizontally into a carrier which is itself then moved downwards into the machine. This action also opens the hinged front flap of the cassette. A pair of posts penetrate into the cassette shell (see Figure 17.6a) and when the thread mode is initiated they move away from the cassette, drawing out a loop of tape in 'M' formation. At the limit of their travel, the posts locate in 'V' notches mounted vertically at each side of the head drum to give the required 186° head wrap. This completes the threading operation, and the posts now form guide rollers on the tape path, as can be seen in Figure 17.6b.

An alternative arrangement is used in some types of VHS decks, notably those designed by Philips. Here (Figure 17.7) the cassette closes over four tape guides and a small pinch roller, all initially positioned *inside* the tape loop across the front of the cassette shell. When threading commences they move away from the cassette in opposite directions. The entry guide, fixed to a loading ring, moves to

Figure 17.6 *VHS threading system: (a) shows start of the threading and (b) threading completed*

the left and then in a loop around the head drum so that the tape embraces the drum in a 'U'-wrap. The outward path of the tape is held clear by two additional posts which move into position as threading progresses. The guide locates in a 'V'-notch at the far side of the head drum to define the tape path at the entry-point of the tape wrap.

Meanwhile the pinch roller, mounted on a toothed rack, moves away from the cassette to the right, taking a loop of tape with it. It comes to rest against the capstan shaft, upon which it is pressed by a spring-loaded lever, again brought into position by the loading mechanism. In this configuration the audio head stack (see later) is positioned directly in front of the cassette.

Tape path

For 'homebase' VHS VTRs, the mechanical and electrical components on the tape deck do not vary fundamentally between any of the formats or manufacturers, with the exception of the Philips deck just described. Let's follow the passage of the tape on its involved journey from the supply spool to the take-up spool. Figures 17.8 and 17.9 show a typical deck, with the tape threaded and ready to go. The supply spool is on the left-hand side. On emergence from the cassette, the tape tension is checked by the tension arm, which brakes the supply spool to

Figure 17.7 *U-wrap tape threading and path in domestic VHS VTR*

Figure 17.8 *The path of the tape through the deck*

Figure 17.9 *Tape path and 'linear' sequence of deck components*

maintain tape tension constant. On most machines this is a purely mechanical operation, though some designs use electrical circuits in a *tension servo*. The tape next encounters the full-width erase head, which in record mode is energised with a high-frequency, high-amplitude CW signal to wipe clean all recorded tracks, video, sound and control. Its mode of operation is the same as the erase head of an audio machine. From here, in some machines, the tape passes around a *supply impedance roller* to iron out any speed or angle fluctuations imparted by the tension arm or the tape's passage over the erase head. An entry guide roller, part of the threading assembly, passes the tape to the slant pole. The latter is inclined from the vertical, and the effect of this is to create more tension on the top edge of the tape than at the bottom. The result is that the tape ribbon is 'nudged' downwards so that it sits firmly on the critically-positioned ruler edge around the video head drum.

The head drum itself is tilted at an angle of about 5° so that the head-wrap angle of the tape is correct with the tape moving parallel to the horizontal deck surface. On leaving the head drum the tape is again nudged downwards by an inclined exit guide to ensure that it is correctly seated on the ruler edge all the way round the 186° wrap of the head drum assembly. The tape now passes over the exit guide roller/threading post and may then be steadied by a take-up impedance roller. Now the tape encounters the audio/sync head, a single assembly with separate heads lined up with the upper and lower edge of the tape. A further roller keeps the tape aligned on its way to the capstan, the prime mover of the tape transport system.

The capstan is a precision-machined shaft, motor-driven under the influence of a servo system and a stable timing reference. Holding the tape tightly in contact with the capstan is the pinch roller, which is disengaged during stop, pause, and fast transport modes. Finally the tape passes back into the cassette and on to the take-up spool, which is gently driven by a slipping clutch or direct-drive motor.

The sequence of deck components is illustrated in the 'linear' diagram at the top of Figure 17.9. In a given format, all models by all manufacturers will conform to this 'linear' diagram so that tapes recorded on any machine will play back on any other. This compatibility is an essential feature, demanding that all machines agree with the parameters laid down in the format specification. The most critical and significant parameters are: video head positioning with respect to height and angular mounting on the head drum; head-to-ruler edge angle on the video head drum; drum exit to sound/sync head spacing; linear tape speed; and of course the track configuration, which we will shortly examine in more detail.

Half-loading

Many VTRs have facilities for real-time counting and index-search systems, which are required to operate in fast-forward and rewind modes. In those machines which do not perform these fast transport functions with the tape fully laced around the head drum, it is necessary to draw a small loop of tape out of the cassette and keep it stretched tight across the audio/control head face so that the control track can be read continuously. It's achieved by one or two *half-loading* arms which, like the guides of Figure 17.6, penetrate the cassette, behind the tape ribbon, as it is lowered onto the deck. During fast-transport mode the half-loading arm(s) moves out from the cassette to route the tape around the ACE head as shown in Figure 17.10. The working principle of real-time counting and tape indexing will be described in Chapter 22.

Figure 17.10 *The half-loading arm (centre of picture) maintains control-track readout during fast tape transport. In FF and REW modes the tape is not laced around the head drum as shown here*

Figure 17.11 *System control sensors on typical VHS tape deck*

Other deck components

Before we leave the deck, however, there are several other devices to mention, mainly concerned with the *systems control* or safety aspect of the VTR. The most important of these are shown in Figure 17.11 starting with the end-of-tape sensors (AA). These prevent physical damage to the tape by inhibiting transport when the end of the tape is reached in either direction, and permitting new motion only in the 'safe' direction. The sensors are optical, depending on a centrally-mounted filament bulb (early models) or infra-red LED emitter, B, to illuminate photo-electric sensors through clear leader tape at each end of the tape. In some front-loading machines the passage of the cassette itself through these light paths is used in place of 'D' as a 'cassette-in' indicator.

Under the cassette carriage there are two microswitches, C and D. One is the 'cassette in' indicator which inhibits mechanical action in the absence of a cassette. The other is a 'tab detector' which prevents accidental erasure if the safety tab has been removed from the cassette in question. More details on these will be found in Chapter 22. Sensor E detects spool rotation on the take-up side. If it ceases, slack tape can build up, so the control system shuts down the deck. Some VTRs have a similar sensor on the supply reel.

Scanning systems

In the original VTR plan, a guard band was left between video tracks on the tape. This was true of the first machine to appear on the domestic scene, the Philips N1500. Linear tape speed here was over 14 cm/second, and the track configuration is shown in Figure 17.12. It can be seen that each video track is spaced from

Figure 17.12 *Guard-band recording: the track formation for VCR format as recorded by the Philips N1500 machine*

its neighbours by an empty guard-band, so that if slight mistracking should occur, crosstalk between tracks could not take place. Each video track was 130 µm wide and the intervening guard bands 57 µm wide. This represents relatively low packing density of information on the tape, and it was soon realised that provided the tape itself was up to it, a thinner head could be used to write narrow tracks; if the linear tape speed was also slowed down the tracks could be packed closer together, eliminating the guard band. Using both ideas, tape playing time for a given spool size could be doubled or trebled. First, though, the problem of crosstalk had to be solved. Even if the mechanical problems in the way of perfect tracking could be overcome so that each head always scanned down the middle of its intended track, crosstalk would occur due to the influence of adjacent tracks, and the effect on the reproduced picture would be intolerable. A solution to this problem was found in the form of azimuth recording.

The azimuth technique

For good reproduction from a tape system it is essential that the angle of the head gap on replay is exactly the same as was present on record, with respect to the plane in which the tape is moving. In an audio system the head gap is exactly vertical and at 90° to the direction of tape travel. If either the record or replay head gap is tilted away from the vertical, even by a very small amount, tremendous signal losses occur at high and medium frequencies, the cut-off point travelling further down the frequency spectrum as the head tilt or *azimuth error* is increased. If the same head is used for record and replay (as is usually the case in audio tape recorders) the azimuth error will not be noticed, because there is no azimuth difference between record and replay systems. A pre-recorded tape from Granny in Scotland will not be up to much, however!

This azimuth loss effect, bad as it is for Granny, is the key to successful recording and replay of video signals without a guard band. Let's designate our video heads A and B, and skew A's head gap 15° clockwise and B's head gap 15° anticlockwise as in Figure 17.13a. This imparts a total 30° difference in azimuth angle between the two heads, and the result is video tracks on the tape like those in Figure 17.13b.

With the built-in error of 30°, head A will read virtually none of head B's tracks, and therefore the guard band can be eliminated. This was the *modus operandi* of the VCR-LP format, using the same cassette and virtually the same deck layout as

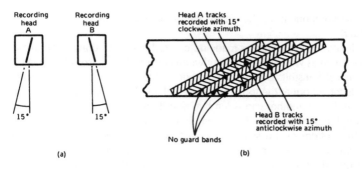

Figure 17.13 *Azimuth offset. The head gaps are cut with complementary azimuth angles so that the guard bands of Figure 17.12 can be eliminated*

the original VCR format, but with linear tape speed reduced by 50 per cent, and video track width down to 85 μm. It worked, and the two-hour machine was a reality.

Subsequent formats use a smaller azimuth tilt: 6° for VHS and 10° for Video 8. The offset *between* heads is double this figure in each case.

Compatibility

The format specification sets out the parameters of the recorded signal on the tape and it is important that each machine in record and playback conforms exactly to these parameters. If a VTR is in a worn or maladjusted state, electrically or mechanically, it may well record and play back its own tapes satisfactorily, but mistracking and other problems will arise when it is called upon to replay tapes from another machine, or when its tapes are replayed elsewhere. Possibly the most common cause of incompatibility in a VTR is mechanical misalignment of deck components, particularly head-drum entry/exit guides, dirt build-up the head drum ruler-edge guide and physical displacement of stationary heads and pinch roller.

To check mechanical and electrical alignment a precision test tape is used, recorded on a 'design centre' machine under carefully controlled conditions. Also required is a series of mechanical jigs produced by the VTR manufacturer, as described in Chapter 25. A properly aligned machine will be compatible with others of the same format, needing at most an adjustment of the tracking control during replay. The function and need for this control will be described in Chapter 21.

Television recap

Before we go on to describe the laying-down of the video tracks in detail, let us briefly recap on the characteristics of the UK television system, known as CCIR system I. For the purposes of this chapter we need only examine the luminance, or monochrome component, of the signal; the colouring signals will come under scrutiny later!

As is well known, the picture is made up of 625 lines, each drawn from left to right of the screen as we view it. For bandwidth conservation reasons (see page 2) the lines are not transmitted in sequence, but in interlaced fashion. This means that line 1 is traced out at the top of the screen, then a gap is left before drawing line 2. Below line 2 another gap is left between it and line 3, and so on, all the way down to the bottom of the TV screen. By the middle of line 313, we are the centre of the bottom of the screen, and at this point the scanning process is suddenly terminated, recommencing at the top of the screen. So far we have traced out one *field* of $312\frac{1}{2}$ lines in a period of one fiftieth second or 20 ms. During this time the instantaneous brightness of the scanning spot along each line has been changing in sympathy with the video signal to build up the picture. The scanning lines for the first field are shown in solid line in Figure 17.14.

The lines of the second field (nos 313 to 625) are slotted into the gaps between the lines of the first field as the vertical timebase commences its second journey down the screen, shown by the dotted lines of Figure 17.14. The second field lasts another 20 ms, and the two fields combined make up a television *frame,* one of which is completed every one twenty fifth of a second, or at 40 ms intervals. Each field contains $312\frac{1}{2}$ lines, and the duration of each line is 64 µs.

The television waveform contains two sorts of synchronising pulses, one at 64 µs intervals to define the starting point of a new line and a more complex one at 20 ms intervals to signify the beginning of a new field. These are illustrated in Figure 17.15a and b respectively, and are separated from the vision signal in the TV receiver to initiate the flyback or retrace stroke of the line and field timebases. Thus the video signal is 'chopped' as it were, at line and field rate, and a sharply-defined pulse inserted as a timing reference. In a VTR the timing reference at field rate is a useful marker, and is used to set the video head position and define the length of a video track on the tape. We can now relate the TV picture to the magnetic 'signature' it writes on the video tape.

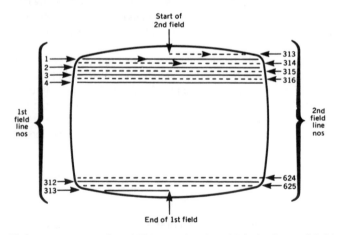

Figure 17.14 *Interlaced TV scanning, in which the lines of field 2 are traced in the gaps between those of field 1*

234

Figure 17.15 *Line and field synchronising pulses in the television waveform*

The video track

Figure 17.16 shows four adjacent tracks on the tape. Track 1 is laid down by head A, and the 'phasing' of the spinning video head drum is arranged so that the A head enters on to the tape and starts to record just before a field sync pulse arrives. It will write about seven lines of picture before recording the field sync pulse, and then go on to write the rest of the lines in the field. By the time line 306 has been recorded, head A is leaving the top of the tape, having recorded one field of $312\frac{1}{2}$ lines; head B has entered onto the tape and is about to record track 2, consisting of the next sync pulse and field; and the head drum has turned through half a revolution, or 180°. During this time the capstan has pulled the tape through the machine just far enough to ensure that track 2 lays alongside, and just touching, track 1. Track 3 is laid down by head A again, track 4 by head B, and so on.

We can see, then, that signals recorded towards the lower edge of the tape correspond to those in the top half of the picture and *vice-versa*. Thus a tape damaged by creasing along the top edge may be expected to give a horizontal band of disturbance in the bottom half of the reproduced picture. Around the period of the head changeover point, both heads are at work for a brief instant, one just about to run off the top of the tape, and the other having just entered at

Figure 17.16 *The positions of TV lines and fields in the recorded tape track. Each head writes one field of video information*

235

the bottom. Thus there is an overlap of information. All VTRs incorporate a head switch which electronically switches between the heads at the appropriate time, just before the field sync pulse. As should now be clear, this takes place at the very bottom of the picture, and any picture disturbance due to head changeover during record and replay is hidden by the slight vertical over-scanning which takes place in a correctly-adjusted TV. Figure 17.17 shows the head changeover on a TV whose picture height has been reduced to demonstrate the effect.

Figure 17.17 *The video-head switchover period generates a picture disturbance over the bottom few lines of the tape-replay picture. Normally hidden below the visible image, here it is revealed by adjustment of the monitor's picture height control*

Still-frame considerations

In domestic VTRs the video head drum and the tape move in the same direction, so that with the head drum spinning anticlockwise (as they do in all formats) the tape is pulled through the half-wrap round the head drum in an anticlockwise sense. This, in conjunction with a special lapped surface on the rim of the head drum, ensures a minimum of friction between tape and drum.

From our studies so far, and the video track diagrams already described, it might reasonably be supposed that if the tape transport were stopped while video head rotation continued, the heads would repeatedly scan the same track over and over again to produce a good still-frame picture. In fact it does not happen, for reasons which will become clear.

Referring again to Figure 17.9, the angle between the plane of the head path and the ruler edge around the head-drum mounting is 5.032° (VHS system) and when the tape is stationary the video heads move across the tape at this precise angle. Once the tape is moving, however, the tape and head velocities subtract,

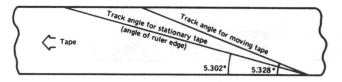

Figure 17.18 *The effective video track angle changes when tape transport is stopped for reasons explained in the text*

because the head and tape are moving in the *same* direction. Think of the tape as a slow lorry being overtaken by a fast car representing the video head; the car's speed *relative to the lorry* is less than if the lorry were stationary. This effective slowing of the writing speed means that in the fixed time available (20 ms, one half-turn of the head drum) the track angle will be steeper than the 5.302° set by the guide, and the recorded track length shorter than when the tape is stationary. The relative track angles and lengths for stationary and moving tape are shown in Figure 17.18, where it can be seen that the track angle has changed to 5.328° when the tape is moving in record and replay. Although the angles and parameters quoted are for VHS, the same principle applies to all formats. Let's see what effect this has on still-frame reproduction.

The video track, recorded at an angle of 5.328°, is now stationary and being scanned by a head moving across the tape at 5.302°. The difference is not very great, but with video track width of the order of 49 μm it is sufficient to cause the head to diverge from its intended track on to an adjacent one, which, as we have already seen, is recorded with the 'wrong' azimuth angle. This results in mistracking and a band of noise on the reproduced picture, as shown in Figure 17.19. Where the noise band occurs depends entirely on the precise stopping point of

Figure 17.19 *Head-mistracking effect on playback picture*

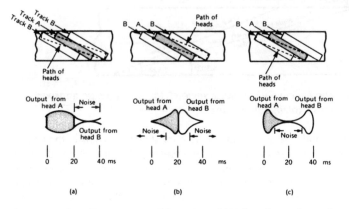

(a)   (b)   (c)

Figure 17.20 *The position of the mistracking bar depends on the exact point, relative to the heads' path, at which the tape stops*

the tape. Figure 17.20 shows three possibilities, with their effects on the still-frame picture. In simple VTRs where a stop-motion facility is provided, this unavoidable noise band precludes any serious use of the feature, and it is referred to as 'picture pause' rather than 'still frame.'

Picture freeze

Second-generation VTRs had special facilities for the display of noise-free still frames. First, let's dismiss what appears to be the simple solution, that of altering the angle of the head ruler edge during still-frame replay. It would appear that slightly tilting the ruler edge or inclined tape guide around the head drum mounting would realign the tracks with the head path during still-frame reproduction, and solve the problem. The guide is precision-machined into the lower head drum with great accuracy, however, and cannot be altered. Even if the inclined tape guide could be somehow made to move slightly on the lower drum, it would be mechanically impossible to arrange a system whereby it was physically tilted through the tiny angle involved with any useful degree of accuracy and repeatability. Other solutions have to be found!

One approach is to make the video heads wider, so that in the stop-motion mode the replay head is 'reading' a sufficiently broad path to embrace the change in track angle without undue loss of signal. The idea is shown in Figure 17.21 where a head 59 μm wide is able to keep in sight the 49 μm video track throughout the field period. It must be appreciated that in stop motion both heads trace the *same path* across the tape, so that if track 1 (head A's province) is being scanned, head B will produce no output from it due to the azimuth error (refer again to Figure 17.20). To obtain a useful output from head B, then, it needs to see some of its own track 2, and to this end it is made even wider than head A, in fact 79 μm. Even when travelling along the same path as head A, then, it reads enough of the adjacent track 2 to provide a usable output, though the picture quality in

Figure 17.21 *A special wide video head scanning a video tape track during 'stop motion'*

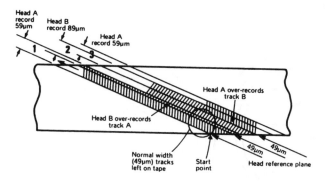

Figure 17.22 *Laying down standard-width tracks with wide recording heads. The excess track width is 'wiped off' by the next head sweep*

still frame is, not surprisingly, inferior to that on normal replay. The extra-wide heads are also necessarily used during record to lay down normal compatible 49 μm tracks according to the format specification. This is achieved because the lower edges of both heads are on the same reference plane, and each new track recorded by either head will wipe off, or over-record the excess width of its predecessor, as shown in Figure 17.22.

Three-head and four-head drums were introduced to improve still-frame reproduction. The extra heads are wide, and optimised for 'freeze' reproduction; a further advantage was that both 'trick' heads could now be cut with an azimuth angle to scan *the same track*. Figure 17.23 shows one such approach. Electronic field-store memories can be used for consummate freeze-frame pictures in home VTRs.

Miniature VHS head drum

The 62.5 mm drum diameter of the standard VHS specification is a great handicap in portable video equipment. To achieve a deck size small enough to be accommodated in a light camcorder a small head drum is used: it is 41.3 mm in diameter. To permit record and playback of *standard* VHS tracks some complexity in

Figure 17.23 *One form of three-head drum. In record and normal playback heads A and B are used. For still-frame, heads B and B' scan the same video track for noise-free and jitter-free reproduction*

Figure 17.24 *Small VHS head-drum. Four heads are required to read and write standard tracks*

in the mechanics and electronics of the machine is unavoidable, and Figure 17.24 shows the essence of the arrangement. The travel of the tape guides follows a longer and more sinuous path than before, to wrap the tape around 270° of the periphery of the small drum, which rotates at 2250 rpm. This high speed is calculated to sweep a single drum-mounted head along the entire length of a *standard* VHS video track during its contact with the wrapped tape. The inclined and continuous ruler-edge around the lower drum assembly maintains the tape at the normal 5.302° angle to the head-sweep path.

Plainly, one pair of heads will not suffice to work this system. At the point when one head is leaving the tape wrap, there needs to be another just 90° ahead, the point where it is just entering the tape wrap. This ensures continuity of signal feed onto the tape, whose linear speed around the drum conforms to standard VHS specifications – 2.34 cm/second for SP mode, 1.17 cm/second for LP mode. If the head which has just left the tape is writing or reading 'A' tracks the one ahead of it

Figure 17.25 *Sequential switching for the heads in Figure 17.24. Switching is applied on record and playback*

and the one behind it must be 'B' heads, with azimuth angles cut accordingly. Hence the A-B-A-B configuration of the four heads around the drum in Figure 17.24. Each head scans every *fourth* track on the format-standard tape.

At any given moment only one of the four heads will be active in record or playback, and since two others will be in contact with the tape at this time a four-phase head-switching system is required during both record and replay. Figure 17.25 shows the switching sequence and the time relationship of the video signal to the active period of each head. The switching system is the same in record and playback modes, though of course the routing of the video FM carrier is opposite.

It can be seen, then, that four heads are necessary in a 'small' VHS head drum to do the work of the two in a conventional drum. If separate heads are provided for SP and LP eight heads are required on the drum, though they can be mounted in four chips, each carrying two windings and two head gaps. Some VHS camcorders are additionally fitted with a flying erase head, giving an effective total of *nine* heads around the drum periphery, with a multi-winding rotary transformer to couple the recording signals to the heads.

Video-8 format was designed from the outset for a two-head 40 mm-diameter drum, so these complications do not arise.

Tracking

We have seen that during record the phasing of the head drum position relative to the incoming video ensures that the field synchronsing pulse is laid down at the beginning of each head scan of the tape. During replay it is necessary to set and maintain the relative positions of the tape and head drum so that each head scans down the centre of its own track. The record phasing and replay tracking is carried out by the VTR's servo systems, which have a chapter to themselves later in this book. Suffice it here to say that on record a *control track* is recorded along the edge of the tape as a timing reference. This takes the form of a 25 Hz pulse train and is used on replay to set and maintain the relative position of the tape tracks and head drum.

Automatic tracking systems

The system of marking the physical position of each video track on tape by control-track signals is a long-established one, and lends itself to various methods of quick programme finding by modifying the control track pulses temporarily at the beginning of each new recording. The VISS and VASS search systems used in VHS machines detect the change of control-track pulse formation in search, FF and REW modes to stop transport. More of this in Chapter 22.

Servo control by a separate tape track, however, requires provision and careful alignment of a separate pulse record/replay head; and a tracking control which in many cases is a manual (user-operated) type. Correct tracking with this arrangement is also dependent on correct tape tension and precision alignment of tape guides. For these reasons and others, control-track systems are vulnerable to tracking errors, especially in LP modes where the video tracks on tape are narrow and less tolerant of head-path errors.

Automatic track finding

Automatic track finding (ATF) is a more advanced form of tracking, in which head-guidance signals are continuously recorded *in the video tracks*. The concept was first introduced in the now-defunct Philips/Grundig V2000 format, where it was known as DTF (dynamic track following) and was used with a piezo-bar mounting system for the heads, whose position could be set by applying DC deflection voltages to the piezo bars.

Video-8 format uses an ATF system, the essence of which is a pilot tone which is recorded with the picture throughout every video track. Four pilot tone frequencies are used: f1, 101.02 kHz; f2, 117.19 kHz; f3, 162.76 kHz; and f4, 146.48 kHz. They are added to the luminance FM record signal and recorded in the sequence f1, f2, f3, f4 in successive head sweeps, see Figure 17.26. Relatively low frequencies like these are almost unaffected by any azimuth offset of the replay head, so pilot-tone crosstalk from adjacent tracks is easily picked up by the video heads during playback. The pilot-tone frequencies are chosen to have specific relationships as Figure 17.27 shows. The beat frequencies which arise when pilot tones from adjacent tracks are mixed are always 16 kHz or 45 kHz. These beat products are used to steer the head path/tape track alignment for optimum tracking: when the levels of 16 kHz and 45 kHz beat signal are equal the replay head must be scanning along the dead centre of its video track, indicating optimum tracking.

There are several ways in which the ATF pilot tones can be processed during playback. A simple one, illustrated in Figure 17.28, utilises two bandpass filters to

Tape movement ← f1 f2 f3 f4 f1 f2

Head scan

Figure 17.26 *Tone sequence laid on tape for ATF*

Approx 45 kHz — f1 101·024 kHz — Approx 16 kHz

f4 146·484 kHz — f2 117·188 kHz

Approx 16 kHz — f3 162·760 kHz — Approx 45 kHz

Figure 17.27 *ATF tones have a carefully-selected frequency relationship*

pick off and separate the 16 kHz and 45 kHz beat products so that they can be separately detected and measured. The DC outputs are applied to the differential inputs of an operational amplifier whose output forms the error signal. This error output can be used to phase-lock either the capstan or head-drum servo to give accurate and continuous tracking correction with built-in compensation for 'mechanical' errors, tape-stretching etc.

ATF playback

While Figure 17.28 shows a simplistic approach to replay ATF processing, the actual system used is more sophisticated. Pilot tones are generated during play-back, again changing on a track-sequential basis triggered by head tacho pulses. This time the tones are sequenced in reverse order, however: see Figure 17.29. The top waveform represents the head-drum flip-flop signal, high for head A, low for head B. The off-tape pilot tone (REC pilot) is shown below, and at the bottom the newly-generated local pilot tone (REF pilot) in the reverse order f4, f3, f2, f1. The diagram shows ideal tracking conditions, in which the off-tape tones switch in synchronism with the REF pilot tones. At every fourth field two f1 tones appear simultaneously at a mixer, whose output consequently drops to zero. During the other (properly-tracked) fields the beat product is either zero (f3/f3) or 29 kHZ (f2/f4), the latter being outside the band of interest and thus rejected.

If the tape speeds up – see Figures 17.26 and 17.29 – the REC pilot moves to the left, permitting REC f2 to appear during the period of REF f1 and giving rise to a 16 kHz beat product. As other tracks are scanned all the beat products (f3/f4, f4/f3, f1/f2) are 16 kHz.

Conversely, when the tape slows down the REC pilot pattern moves to the right. Some REC f4 now appears during the REF f1 period to produce a 45 kHz beat

Figure 17.28 *Basic arrangement for derivation of an error voltage from inter-track pilot beats*

Figure 17.29 *Replay beat-tone generation: off-tape pilot frequencies are compared with a locally-generated tone sequence*

product. Similarly during subsequent scans REC f1 beats with REF f4, f2 with f3 and f3 with f2, producing a 45 kHz beat output in each case.

Thus a fast-running tape always results in a 16 kHz output from the mixer; and a slow-running tape always produces a 45 kHz output. By using suitably tuned bandpass filters to select these products, and feeding their outputs to separate peak-detectors an ATF error signal is produced for subsequent smoothing and passage to the capstan servo control input. In practice the replay ATF processing circuit is also provided with artifices to detect 'false lock' conditions and to permit locking of mistracking-noise bars during search modes.

Video-8 tape-signal spectrum

The DTF pilot tones are recorded at the lowest part of the frequency spectrum as shown on the left of Figure 17.30. Here they do not interfere with the 'signal' components of the tape recording. The other parts of the tape-signal spectrum will be dealt with later: luminance in Chapter 18, chroma in Chapter 19 and audio in Chapter 20.

Figure 17.30 *The spectrum of signals on tape – Video-8 format. That for Hi-8 format is given in Chapter 26*

18 Signal processing: luminance

In previous chapters, we have seen some of the problems of tape recording video signals, and have examined how these are overcome by the use of high writing speeds and FM modulation. We have seen, too, how home VTRs achieve an acceptable performance at low cost by signal bandwidth restriction. This is by no means the only shortcoming of the domestic machine! In spite of these, the playback picture from a domestic VTR in good condition is very acceptable; in this chapter and the next we will examine not only the basic circuitry used to process TV signals within the VCR, but also the special circuits and processses, some of which owe their origins to broadcast technology, which compensate for the deficiencies inherent in a cassette VTR system, and enhance performance.

The luminance (or black-and-white) signal is dealt with separately from the chrominance (or colouring) signals in home VTRs. For luminance the basic idea is to modulate the signal on to an FM carrier for application to the recording head, and demodulate it to baseband during the replay process. As the simplified block diagram of Figure 18.1 shows, however, there are several other processes undergone by the luminance signal, and these will be described in turn.

Figure 18.1 *The stages in the luminance-recording process*

AGC (automatic gain control) and bandwidth limiting

The basic luminance signal that we wish to record may come from a TV camera or other local video source, or more likely, a broadcast receiver built into the VTR. In either case it will be positive-going for white and will probably contain a chrominance signal modulated on to a 4.43 MHz carrier. It is important that the signal recorded on the tape is within the limits of the recording system, so the luminance signal is first passed through an AGC amplifier with a sufficiently wide range to compensate for signal inputs of varying amplitudes. This works in a similar manner to the AGC system of a radio or TV, by sampling its output level to produce a

DC control potential, and applying this to an attenuator at the amplifier input. Thus the output from the AGC stage will be at constant (say) 1 V amplitude.

In home VTRs, luminance bandwidth is restricted to about 3 MHz, give or take a hundred kHz or so between the formats. If higher frequencies than this are allowed to reach the modulator they will make mischief with sidebands, as explained in Chapter 16. A low-pass filter with a quite sharp cut-off around 3 MHz is incorporated in the record signal path, then, and this also eliminates all the chrominance components of the signal, which are based on a subcarrier of 4.43 MHz. When recording in monochrome more bandwidth can be allowed, occupying the space normally reserved for the chrominance signal, and many machines have an automatically-switched filter characteristic for monochrome and colour recordings, as shown in Figure 18.2.

Figure 18.2 *VTR luminance-recording filter characteristics relative to the spectrum of the composite video signal as broadcast*

Pre-emphasis

Chapter 16 showed how pre-emphasis is used in an FM system to reduce noise. What is required is a boosting of high-frequency (HF) signals before the FM modulator, and this is achieved by a circuit like that in Figure 18.3a. Here we have a common-emitter transistor with a small capacitor (C1) in its emitter circuit. At high frequencies, the capacitor's reactance will become comparable with the emitter resistor R1, and the negative feedback due to the latter component will diminish, resulting in greater output from the stage (i.e. at the collector) at those frequencies. A typical pre-emphasis curve is shown in Figure 18.4a, in which 10 dB of pre-emphasis is given to frequencies above 1 MHz.

In some machines, particularly Video-8 types (which make use of very narrow video tracks) a further, *non-linear*, pre-emphasis circuit is used. This applies a degree of HF boost which is dependent on signal level (greater for small signals, less for large signals). A family of curves for this system is given in Figure 18.4b. Although this calls for more complex circuits (the de-emphasis characteristic has to be non-linear to compensate) it does offer a useful reduction in noise level on replay.

(a)   (b)

Figure 18.3 (a) Rudiments of pre-emphasis circuit, and (b) its effect on the luminance staircase waveform

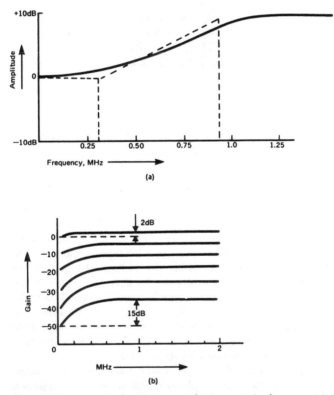

Figure 18.4 A typical VTR pre-emphasis curve is shown at (a). In (b) appears a family of curves for a non-linear pre-emphasis system, in which the degree of HF 'lift' varies from 2 to 15 dB, depending on signal level

Because the steep rising and falling edges of a luminance transient, or step, look like a high frequency they are effectively differentiated in the pre-emphasis process, so that the 'risers' of a luminance staircase waveform take on the spiky characteristic shown in Figure 18.3b.

Where the spikes exceed peak white level as defined by the output of the AGC stage, we are once again in danger of driving the FM modulator into excessive deviation, and this must be prevented.

White and dark clip

The next process, then, is a clipping action on the video signal. The danger areas for overmodulation are those shown at points X and Y in Figure 18.3b, and the dotted lines indicate the levels at which the signal must be clipped. This action takes place in a circuit like the one of Figure 18.5, where diodes provide the clipping effect at both extremities of the the luminance signal. Let's assume the sync-tip of a normal luminance waveform drives Tr1's emitter down to +1V, and peak white takes it up to +4V. If RV1 is adjusted so that +1V is present at D1 anode, then ignoring the voltage drop access the diode itself, D1 will conduct on any luminance signal excursion below 1V and the signal will be grounded in C1.

Similarly, RV2 is set to give +4V at D2 cathode so that any luminance signal above this level turns on D2 and is grounded in C2. Thus RV1 becomes the 'dark clip' level control and RV2 the white clip level control. In practice, a little 'headroom' is permitted above and below the normal signal excursion before the clippers come into operation; some machines have only a white clip facility.

For this clipping system to work, the voltage levels corresponding to black and permissible peak white have to be closely defined. Because the AGC stage defines the overall signal amplitude, it is only necessary to clamp the black level to a fixed voltage to achieve this, and clamping is carried out immediately prior to the clipper stage, usually in a driven line-rate clamp similar to those used in TV receivers.

Figure 18.5 *A diode threshold clipper; clipping levels are set by RV1 and RV2*

FM modulator

The modulator is one of the most critical stages in a VTR. Its output needs to have equal mark-space radio regardless of deviation, and deviation must be linear with respect to modulating (video) voltage. The output must not contain harmonics which would generate harmful sidebands, and frequency stability has to be of a high order. A constant output level is also required over the whole of the deviation range, but this can be achieved by limiting stages after modulation, as we shall see. Current VTRs use an IC modulator, but to explain the operation of the circuit we shall take as a model the astable, or multivibrator, type as used in early VTR designs. The curve of Figure 18.6 shows the relationship between modulating voltage and output frequency for a typical system.

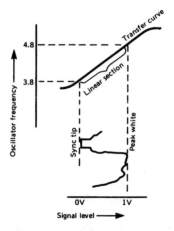

Figure 18.6 *The transfer curve for an FM modulator. It has to be very linear between the frequencies corresponding to sync-tip and peak white. For the VHS system illustrated, these are 3.8 to 4.8 MHz*

Astable modulator

The astable oscillator is a well-known configuration and is drawn in a form suitable for VTR use in Figure 18.7. The square-wave output is produced by alternate conduction in Tr1 and Tr2, the conduction period for each transistor being set by the capacitor/resistor (CR) combination in its base circuit. Thus C1/R1 determine the conduction period of Tr1; when Tr1 switches off Tr2 comes on for a period determined by C2/R2, and so on. The time-constant of these CR pairs is chosen to be identical, i.e. C1=C2 and R1=R2 so that the *mark-space ratio* of the output waveform is equal or 1:1. Now the conduction period for each transistor is also dependent on the *aiming voltage* of each of the timing capacitors, that is the voltage towards which it is charging via its timing resistor. If this aiming voltage is varied, the charging *rate* of the cross-coupling capacitors C1 and C2 varies in

Figure 18.7 *An astable multivibrator for luminance FM modulation*

sympathy, and it is the charging *rate* of the capacitors which determines the output frequency. In the circuit of Figure 18.7 the aiming voltage is set by the potential divider R3/R4, but this can be modified by an externally applied voltage at point X. Because this is common to both CR timers, the mark-space ratio will remain constant while the basic frequency is varied in proportion to the voltage applied to point X. This, then, is how voltage variations (the video signal) are converted to frequency variations in the output of the astable modulator.

We have seen that harmonics are harmful in the output from an FM modulator, and the square-wave output characteristic of an astable circuit is very rich in harmonics! These are lost in the tuned and balanced transformer T1, whose output is virtually sinusoidal.

As frequency deviation is proportional to applied voltage, the carrier frequency and deviation are set by adjusting the video DC level and gain in the preceding circuits. Thus in our previous examples, basic carrier frequency (corresponding to black level) would be set by potentiometer adjustment of the video clamp voltage, and maximum deviation by presetting the operating point, or gain, of the AGC amplifier.

Limiting and head drive

A degree of amplitude modulation can occur in the FM conversion and coupling processes, and it is important that this is ironed out before the signal is applied to

Figure 18.8 *Simple diode amplitude clipper*

the recording heads. Typically, diode clippers are used in a circuit akin to that already given for white and dark clippers. The difference here is that the clipping action is constant, a process known as limiting. In Figure 18.8 the diodes D1 and D2 limit the signal to a constant level for application to the recording amplifier.

The recording amplifier has two basic functions: to provide power amplification to drive the recording heads, and to apply recording equalisation to compensate for the falling response of the video heads and head/tape interface with increasing frequency. The circuit diagram of Figure 18.9 shows the arrangement. The first stage after the limiter, Tr1, provides equalisation by giving a degree of lift to the higher frequencies. Frequency-selective negative feedback is again used, the operative component being C1. An emitter follower stage Tr2 acts as a buffer to drive the complementary output pair Tr3/Tr4. These work in similar fashion to audio and field output stages in radio and TV sets, but at a much higher frequency.

The level of the FM luminance signal fed to the recording heads is critical, and is set at an optimum level for the type of videotape in use. The constant-level carrier acts as a recording bias signal for the colour signal which accompanies it

Figure 18.9 *An FM recording amplifier by Panasonic*

on to the tape, and the critical FM luminance *writing current* is set by VR1 in Figure 18.9 to bias the recording level to the centre of the linear flanks of the transfer characteristic shown in Figure 16.4.

Transferring the video signal to a pair of heads rotating at 1500 rpm is something of a problem, and the later recovery of the tiny playback signal an even greater one! Any form of brushgear is impractical from the noise point of view, so rotating transformers are used, with the primary winding stationary and the secondary winding rotating with the heads. These roles are reversed during playback, of course. Two separate and independent transformers are used, having printed windings and ferrite cores; signal transfer takes place across a very narrow air-gap between the two ferrite discs. Multi-head VTR designs use separate and concentric transformers. Except for small-drum VHS systems (Chapter 17) head switching is not carried out on record, both heads being driven together via the rotary transformers from the output of our push-pull recording output stage (Figure 18.9).

IHQ

'Intelligent high quality' is the name given to a concept developed by Akai for home VTR use. In similar fashion to high-quality audio-cassette recorders, it involves a short test-recording just before the 'real' recording takes place; its purpose is to establish the magnetic characteristics of the tape in use. In a fully-automatic sequence taking a few seconds, a test signal is recorded on the tape, which is then rapidly rewound and replayed for analysis in a measuring circuit. The data thus obtained is memorised and used to set the response of a filter in the luminance record amplifier stage which optimises the shape of the lower sideband recorded on tape. The effect is one of equalisation of the energy in the spectrum of signals on tape, leading to enhanced detail and lower noise levels in the reproduced picture.

Replay circuits

During replay, the off-tape signal from the video heads is very small, and to maintain the necessary >40 dB S/N ratio in the reproduced picture, low-noise amplification is necessary. The main replay processes after the preamplifiers are head switching, equalisation, limiting, drop-out compensation, demodulation to baseband, de-emphasis, crispening and amplification, after which the luminance signal is restored to its original form, usually 1V peak-to-peak, negative-going syncs. A block diagram showing the order of the replay circuits is given in Figure 18.10. We will examine these in turn.

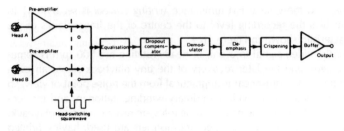

Figure 18.10 *The replay chain for the luminance signal*

Head preamplifier and switching

The RF signal from the rotary transformers contains signal components ranging from 100 kHz to over 8 MHz. These are at high impedance and very vulnerable to noise pick-up. A typical pre-amplifier circuit appears in Figure 18.11. The replay signal is applied to a balanced transformer T1 whose windings resonate with trimmer C1, set to 'peak' the LC circuit at about 5 MHz. This resonance is damped by variable resistor RV1 to provide a reasonably smooth frequency response at the input to Tr1. An FET device is used here, exploiting its characteristics of high input impedance (to minimise loading of the head circuit) and low noise figure. ,

Although both heads are driven together during the record process, on all current formats head switching is carried out on replay. This serves the dual purpose of eliminating noise from the inoperative head (remember only one of

Figure 18.11 *Head pre-amplifier circuit. Tr3 and Tr3a are head-switching transistors, and RV2 is a balance control to equalise the signals from each video head*

the pair is scanning the tape at any given moment) and sharply defining the head changeover point to minimise picture disturbance during changeover. In Figure 18.11 head switching is carried out by Tr3 and Tr3a. An incoming square-wave signal at 25 Hz derived from a *tachogenerator* on the head drum switches Tr3 and Tr3a alternately into saturation, shorting to ground the output of each head during its 'passive' period. The FM replay signal at point X is continuous, then, and sourced from each head alternately for 20 ms periods.

Equalisation

Chapter 16 explained why during replay the signal from the tape head falls at the rate of 6dB per octave. HF losses due to the approach of $f$ex are largely compensated for by the resonance circuit associated with the replay head and described above. To even-out the response on replay, further equalisation is provided, and in practice it takes the form of a 'boost' in the 2–3 MHz region, the major area for the lower FM sideband of the luminance signal. This is often catered for by the provision of an inductive component in the collector circuit of the preamplifier or a following stage.

Limiting

For correct operation of any FM demodulator, it is important that its input signal contains no amplitude variations. For this reason, several stages of limiting are usually provided in the path of the FM replay signal to clip it to a constant level. The form of diode clipper shown in Figure 18.8 (D1 and D2) is suitable for this. In current VTR designs, limiting normally takes place (along with much of the other replay processing) within an IC.

Replay picture optimisation

It is on the playback RF carrier that the Nokia ASO (active sideband optimisation) system works. One cause of poor high-frequency response in a tape replay system is the random phase distortion which occurs at sudden frequency transitions in the off-tape signal, representing vertical edges and outlines of picture objects. The ASO concept depends on a small LC delay network, working in conjunction with a pair of clamp diodes. This L/C/D network has the effect of speeding up the transition between the old and new levels in the demodulated video signal to give 'cleaner' and better defined edges.

The action of the Akai IHQ system (whose function in record mode we met a few paragraphs ago) during replay is on a noise-reduction system, whereby the degree of noise cancellation is made dependent on the S/N level of the replayed luminance signal. This adaptive noise-cancellation scheme optimises playback quality from a wide range of tape types and signal qualities, ranging from a low-noise recording on a high-grade tape to a well-worn library cassette.

Dropout compensation (DOC)

A videotape is not the perfect medium that we would like it to be! The magnetic coating is not completely homogenous, and with video track widths smaller than the diameter of a human hair even a microscopic blemish in the magnetic coating will delete some picture information. As the tape ages, slight contamination by dust and metallic particles, and oxide-shedding effects, will aggravate the situation. The effect of these tiny blemishes is a momentary loss of replay signal known as a *dropout*. Unless dropouts can be 'masked' in some way a disturbing effect will take place in the form of little ragged black or noisy 'holes' in the reproduced picture.

In practice, the video information on one TV line is usually very much like that on the preceding line; so that if we can arrange to fill in any dropout 'holes' with the video signal from the corresponding section of the previous line, the patching job will pass unnoticed. What's required, then, is a delay line capable of storing just one TV line of 64 μs duration, so that whenever a dropout occurs we can switch to the video signal from the previous line until it has passed. For delay-line bandwidth reasons, this is difficult to achieve at video baseband frequencies, so it is carried out on the FM signal before demodulation. To avoid disturbance on the picture the switching has to be very fast, virtually at picture-element rate. One form of DOC circuit is shown in block diagram form in Figure 18.12. The FM replay signal takes three paths, the upper of which is via demodulator 1 and on through the following replay circuits; this is the path taken by the signal under normal circumstances, that is, when no dropouts are present. The bottom path consists of a level detector which monitors the FM signal from the playback heads. When a dropout comes along, the FM signal falls below the detector's threshold level (normally pre-set by a potentiometer) and the detector output falls to zero. This operates a fast diode changeover switch S1, which now selects the middle path via demodulator 2, whose FM input is exactly 64 μs (one TV line period) 'old' due to the delay line, and thus corresponds to the FM replay signal at the same point on the previous line. This signal is held by the switch until the dropout has passed and full FM input is restored at the DOC input, when the detector output reappears and the changeover switch drops back to the output of demodulator 1.

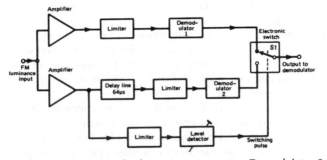

Figure 18.12 *A simple dropout compensator. Demodulator 2 is operating on an FM signal one TV line earlier than that in demodulator 1*

Recycling DOCs

The system outlined above breaks down when any dropout exceeds one TV line duration, and this happens often – the tape area occupied by one line is microscopic. If the dropout period exceeds one line, then, both demodulators in Figure 18.12 will be looking at noise, and a disturbance will be visible on the TV screen. To prevent this, further steps must be taken. In early designs, a 'grey oscillator' was used, consisting of a circuit, active after the first dropout line, which generated a CW signal corresponding to mid-grey in terms of off-tape frequency. The resulting neutral tone inserted in the picture was subjectively less noticeable than the dropout it replaced.

Later designs perform better by recirculating the last 'good' TV line around the delay line and reading it out continuously for the duration of the dropout. This gives better reproduction than the grey oscillator, and a recirculating DOC, based on an IC, is shown in Figure 18.13. The replay FM signal enters the chip at pin 2 and emerges, after amplification, at pin 3. After passing through a switched filter (colour–mono bandwidth switching by transistor X3) the signal re-enters the chip at pin 5 and passes through pole A of the dropout switch to emerge at pin 9, *en route* for the FM demodulator. A second output of the amplified RF signal appears at pin 4, routed through the limiter between pins 11 and 12, then into the dropout detector on pin 14. The dropout switching signal appearing at pin 13 is amplified

Figure 18.13 *A recirculating dropout compensator based on a purpose-designed chip (JVC)*

in the source-follower X8 and reapplied to pin 10 of the chip, whence it operates the changeover switch to select pole B. Here the delayed signal is available, and it is routed not only to the demodulator via IC pin 9, but back to the delay line input for recycling. This circuit requires only one demodulator.

Demodulation

Several methods of FM demodulation are available, and such configurations as the ratio detector and Foster-Seeley discriminator have been used for many years for FM sound recovery in radio and TV receivers. More up-to-date designs use a PLL (phase locked loop) or quadrature demodulator for better performance, but the low modulation index used in the VTR system (operating frequency 3–5 MHz, modulating bandwidth 0–3 MHz) requires a different technique. Two circuit configurations are common in home machines, the delay-line detector and the pulse-counting demodulator.

A circuit for the latter appears in Figure 18.14. The limited FM replay signal, represented by waveform (a) in Figure 18.15, is passed into the circuit at point A. R1 and L1 together form a differentiator, so that only spikes appear at the coupling transformer T1. The timing of the spikes depends on the FM frequency, but the duration of each is the same, set by the time-constant of R1 and L1. This is shown in Figure 18.15b. The secondary windings of T1 feed opposite-polarity spikes to the bases of detector transistors Tr1 and Tr2; Tr1 conducts on positive spikes and Tr2 on negative spikes, so that the frequency is effectively doubled at their common collector load resistor R2, giving rise to the waveform in Figure 18.15c. This frequency doubling, a form of full-wave rectification, greatly assists with filtering after the detector. All that remains, then, is to 'count' the pulses, and this is simply done by an integrator consisting of an RC or LC network, R3 and C2 in Figure 18.14. The time constant is arranged so that C2 is charged on each pulse and discharges through R4 between pulses – thus its average charge is proportional to

Figure 18.14 *Pulse-counting FM demodulator*

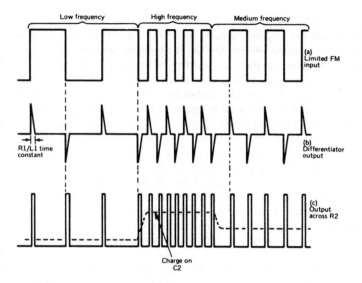

Figure 18.15 *Waveforms for the demodulator of Figure 18.14. The frequency of the incoming squarewave (a) is effectively doubled at (c)*

pulse rate as shown by the dotted line in Figure 18.15c. After further filtering and de-emphasis a facismile of the original waveform is obtained, now at baseband.

The demodulator circuit is carefully designed to offer linear detection throughout the deviation range, and the following filters effectively reject the residual (doubled) carrier ripple. The luminance signal we have now recreated consists of composite video (i.e. with sync pulses present) with a bandwidth of about 3 MHz.

Crispening

The 3 MHz capability of a standard VTR means that replayed pictures will lack the sharpness and definition of an off-air picture; and fine detail, for example the frequency gratings of a test pattern, will not be reproduced. Several other factors are present, though, and these are worth examining. Most video recording involves colour programmes, and colour receivers incorporate notch filters in their luminance amplifiers which impair definition above 4 MHz. At about the same frequency physical limitations of the colour display tube come into play. The pitch of the phosphor-dot matrix on the tube faceplate is such that a picture element *(pixel)* smaller than the area of one RGB dot-group cannot be reproduced, and this is a limiting factor for luminance definition, especially in smaller colour screen sizes. Very little of the content of an average picture consists of fine repetitive detail, and the subjective effect of limited HF response is a blurring of sharp vertical edges.

The crispening circuit goes some way to compensate for HF roll-off by artificially 'sharpening up' vertical edges in the picture. The technique has been used in broadcast and CCTV for many years to compensate for the finite scanning spot size in cameras and FSS systems. In such applications it is known as aperture correction because it has the effect of limiting the scanning spot size to that of a smaller 'aperture' inserted into the tube, and enhancing definition thereby.

The effect of poor HF response is to slow down the rise and fall times of the luminance waveform so that a sudden transition from white to black (or vice-versa) is reproduced as a waveform (a) in Figure 18.16. There are many types of crispening circuit, and a block diagram of an easy-to-understand one appears in Figure 18.17. The luminance step waveform (a) is fed into a short delay line to emerge slightly later at the centre tap, waveform (b). Waveform (a) is also differentiated, producing a negative-going spike, waveform (c). After inversion this spike appears in positive-going form, waveform (d). After a further slight delay waveform (e) appears at the end of the delay line and is differentiated to a spike, waveform (f). Thus we have three carefully timed waveforms (b), (d) and (f) passing into the add matrix. Its output is shown as waveform (g), showing the *pre-shoot* effect of waveform (d) and *overshoot* effect of waveform (f).

The slightly delayed signal now appears to have sharper transitions, and the process works on positive-going steps too. The delay time, represented by the two vertical lines in Figure 18.16, is carefully arranged along with the differentiator time-constants to give optimum correction at a chosen frequency – study of Figure 18.16 shows that above this optimum frequency the crispening circuit actually reduces definition! The degree of pre-shoot and overshoot is usually adjustable by a pre-set control labelled *crispening* or *aperture*.

Figure 18.16 *Waveforms for the crispening circuit of Figure 18.17. Their derivation is explained in the text*

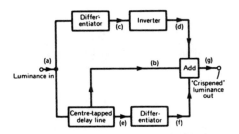

Figure 18.17 *One form of crispening, or aperature-correction circuit*

Luminance delay

The luminance signal processing is now almost complete. Before the playback luminance signal is mixed with the off-tape chrominance signal, however, it is passed through a short delay line. This component serves the same purpose as the luminance delay line in a colour TV, and is necessary because of the differing bandwidths of the luminance and chrominance signal paths. The chrominance signal, being of relatively narrow bandwidth, has a slower rise time than the accompanying luminance signal, and if uncorrected this would have the effect of printing the chrominance component slightly to the right of the luminance image on the TV screen. The delay line, then, slightly retards the timing of the luminance signal to bring the two into step, ensuring correct *registration*.

After amplification, chroma addition, and conversion to low impedance, the vision signal is ready to pass out of the VTR. It is applied to the 'video out' socket in 1V peak-to-peak form, and also to the RF modulator (along with the sound signal) where it is impressed on to a UHF or VHF carrier for injection into the aerial socket of a TV receiver.

S-VHS

The 'super' variant of VHS format overcomes the definition limitations of conventional formats. Here the FM carrier frequency deviates between 5.4 and 7.0 MHz, a 60 per cent expansion. The luminance sidebands are correspondingly wide; the entire luminance FM 'package' occupies a bandwidth of over 8 MHz, compared with 5 MHz in the original plan. S-VHS cassettes have similar running time and appearance to standard types, but contain cobalt-oxide coated tape and can reproduce pictures with 400-line resolution. A diagram of the S-VHS recording spectrum is shown in Figure 26.4.

19 *Signal processing: chrominance*

To understand the colour circuits of a VTR, it is essential to have a working knowledge of the way in which chrominance signals are encoded in ordinary broadcast transmissions. Full details are given in Chapters 3 and 4, and revision of those sections of the book is recommended to readers as a basis for study of the colour section of a VTR. One of the advantages of home VTRs is that at no point is the colour signal demodulated – it remains in subcarrier form throughout the system, giving little opportunity for phase (thus hue) errors. A short recap of the PAL system follows.

PAL colour recap

A TV system requires three simultaneous streams of information to produce a full colour picture; these are the R, G and B (red, green and blue) video signals. For a monochrome signal, such as a black-and-white film, all three are identical and a single signal channel (i.e. the single electron gun and signal path of a monochrome TV set) is all that is required to convey the information. For colour transmissions it is impractical and wasteful to use three separate channels for the RGB signals, and so a method was devised to encode the colouring signals in such a way that they 'ride piggyback', as it were, on the basic luminance signal and the whole can be accommodated within a single 8 MHz-wide transmission channel.

At the sending end the luminance signal is transmitted in a form represented by the waveforms of Figure 17.15. This satisfies a monochrome receiver, and conveys the basic picture – and fine detail – to a colour set. Instead of sending RGB (known as *primary colour*) signals, *colour-difference* signals are produced and these represent, for each primary colour, the difference between the luminance signal and the desired colour at any instant. It is not necessary to send all three colour-difference signals; two will do, because the third can be recreated at the receiver by a matrix fed by the two colour-difference signals we do have, plus the luminance signal. The two colour-difference signals chosen for transmission are those for red and blue, represented by R – Y and B – Y respectively, and each of these amplitude-modulates a *subcarrier* signal at approximately 4.433 MHz. These two amplitude-modulated subcarrier signals, although at the same frequency, are 90° apart in phase, which means that one lags the other by one-quarter of a cycle; they are said to be in *quadrature.* Because *balanced* chrominance modulators are used at the transmitter the subcarrier itself is suppressed, and only the sidebands, or modulation products, appear at the modulator output. The system is called *suppressed carrier modulation,* and depends entirely on the sidebands of the signal to carry the information. Elimination of the carrier (or subcarrier in a colour TV system) has an important advantage. Lying as they do within the luminance passband, any chrominance signals, however coded, will upset the luminance reproduction by causing spurious patterning in the form of dots and lines, and this will happen on monochrome and colour receivers. Careful selection of the subcarrier frequency (4.43361875 MHz, locked to line frequency) minimises the chrominance/luminance crosstalk, and this is further reduced by the suppressed subcarrier technique which ensures that dot-patterns occur only in coloured areas of the picture, and are noticeable only on highly-saturated colours.

Phases and vectors

The subject of phase relationships and vector diagrams is not a widely-understood one. We have already touched on the subject of phase in describing the colour-encoding process, and it will crop up many times in our study of VTR chrominance and servo systems. The word 'phase' describes the timing relationship between two waveforms, usually – but not necessarily – at the same frequency. In Figure 19.1a is shown a sinewave which we will call the *reference; b* shows an identical sinewave with the same timing as a and this is said to be 'in phase' with the reference, as shown by the vector diagram alongside. One complete trip round the vector diagram (360°) represents one cycle of the reference signal, so that waveform *c,* being half a cycle behind the reference signal, is phase-retarded by 180°, or half a circle; it is said to be *anti-phase.* It will be noticed that waveform *c* is an 'upside-down' version of the reference *a,* so that inverting a waveform *if it is symmetrical about the zero line over one cycle* (i.e. a sine-wave or square-wave) will give the same effect as a phase change of 180°. The waveforms involved do not have to be sinusoidal of course; *d* shows a square-wave in phase with the reference and *e* a triangular wave in antiphase, or at 180°.

When two waveforms have a phase relationship of 90° or 270° they are said to be in *quadrature* and their vectors are at right-angles to each other; this quadrature relationship is a useful one, because it means that when the reference waveform is at its peak, its quadrature-companion is passing through zero, so that by 'sampling' at the correct instant we can separate the information carried by each of two mixed signals in quadrature – without crosstalk. This, of course, is the principle used in the synchronous demodulators of colour decoders in TV sets. Waveform *f* in Figure 19.1 is in quadrature with the reference *a.*

263

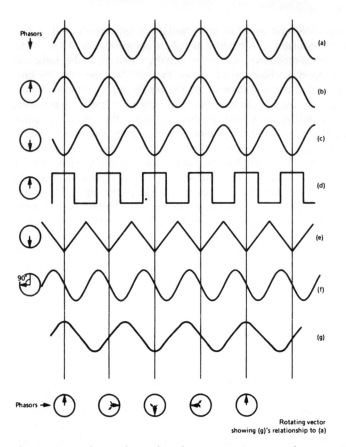

Figure 19.1 *Phase relationships between various waveforms: (a) is the reference and the 'clocks' on the left show vector relationships. Waveform (g) is at a frequency lower than that of the reference (a), giving rise to the clockwise vector rotation shown at the bottom*

Rotating vectors

If we introduce a second signal at a frequency slightly lower than that of our reference, the situation will be as in Figure 19.1g. Here the waveforms start in phase, but the lower frequency (g) will fall behind the reference as time goes on, so its phase lag will increase with time, giving rise to the clockwise-rotating vector shown. After a period, the second signal will be exactly one cycle behind the reference, and the two will be momentarily in phase again, with the vector having completed one full clockwise circle. For a frequency higher than the reference signal, it can be seen that the vector will rotate anticlockwise; in both cases, the speed of rotation of the vector will be proportional to the difference in frequency between the reference and the sample.

If the frequency of the sample is constantly changing with respect to the reference, but by less than one cycle, the result is a 'phase jitter', represented by a
constantly-changing vector angle, but confined to less than 360° (one full turn).
This is what happens, for instance, in a colour TV system, where the reference
signal is a stable subcarrier source, and the 'sample' is the transmitted chroma
signal.

Phase-locked loop (PLL)

A common 'building block' in VTR systems is the phase-locked loop. It takes the
form of an oscillator and phase detector, and is designed to produce an output
signal with a fixed phase relationship to a reference signal; a familiar example is
the flywheel line timebase circuit of a TV receiver.

A simple example of a PLL is shown in Figure 19.2. The process starts with a
phase detector, or discriminator, consisting of a device with two inputs, X and Y.
When the timing of input pulse X is earlier than that of input pulse Y, the detector
will produce a positive DC voltage at its output Z. If, alternatively, pulse Y occurs
before pulse X, output Z will move negatively. When both input pulses are coincident in time (i.e. *in phase*) output Z is zero. Thus the output Z of the phase
detector indicates by its polarity the direction of phase error, and by its amplitude
the amount of phase error. Not surprisingly, this is called the error voltage!

The second half of the PLL is a voltage-controlled oscillator (VCO). This is an
oscillator whose output frequency – and phase – can be controlled by a DC
voltage. It may consist of a crystal oscillator (as in the reference generator of a
colour TV decoder), or an LC or flip-flop oscillator. Its output forms one input to
the phase detector, and its frequency-controlling input is the error voltage from the
phase detector. Thus is the 'loop' set up, and whenever the oscillator and input
signals drift apart in timing, the discriminator generates an error signal to pull the
oscillator back into phase lock with the control pulse, and again achieve zero
phase difference between its X and Y inputs.

The oscillator is then said to be 'slaved' to the reference, or 'master', signal. The
oscillator of a phase-locked loop does not necessarily run at the same frequency
as the control pulse; with a suitable divider in the feedback path from the

Figure 19.2 *Basic operation of a phase-lock-loop*

Figure 19.3 *Phase-lock loop with divider. The output becomes locked to a multiple of the reference frequency*

oscillator to the phase detector, an output locked to a multiple of the 'master' frequency can be obtained, as shown in Figure 19.3.

Quadrature modulation

Addition of the outputs of the R – Y and B – Y modulators at the transmitter renders a single chrominance signal (consisting of just the sidebands of a 4.43 MHz sinewave) which conveys the hue of the colour by its phase, and the saturation of the colour by its amplitude. This chrominance signal is mixed with the luminance signal before transmission, and being based on a frequency of 4.43 MHz it is easily filtered out for decoding within the colour receiver. Figure 19.4 shows the spectrum of the colour-encoded video signal.

A vector diagram, Figure 19.5, puts the colour subcarrier system into perspective. The dotted line represents the phase of the subcarrier signal, and this can extend to any amplitude (length change) to describe saturation; and point in any direction (angle change) to describe hue. Thus the short vector pointing NE illustrated represents a light purple colour; a long vector pointing SW would indicate a bright green colour.

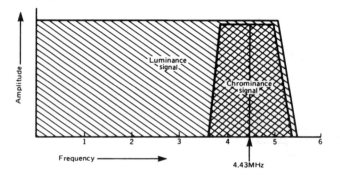

Figure 19.4 *Spectrum of demodulated broadcast video signal*

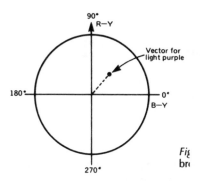

Figure 19.5 *Vector diagram of broadcast colour encoding*

Colour burst

At the receiver we need to recreate the missing subcarrier signal in order to demodulate the chrominance signals. The recreated subcarrier must be in phase with that at the transmitter – phase, as we have seen, means relative timing, and to measure the phase or timing of the colour subcarrier signal, we need a reference against which to do it. A timing reference is transmitted once per line, then, and it takes the form of a 'sample' of ten cycles of plain subcarrier sitting on the back porch of the video signal, illustrated in Figure 19.6 and known as the *colour burst*. This is gated out in the TV's decoder and used to phase-lock a crystal oscillator, thus providing the necessary timing reference.

An essential feature of the PAL system is the effective inversion of the R – Y subcarrier signal on alternate lines. This is done to eliminate hue errors due to phase distortion in the signal path, and facilitates cancellation of such errors in a delay-line matrix circuit. Thus on TV line 1 the R – Y axis is OB in Figure 19.7, whereas on line 2 the R – Y axis is OD. The B – Y axis is OA on all lines, so that magenta (a shade of purple containing equal proportions of red and blue) would appear as vector OX on line 1, OY on line 2, OX on line 3 and so on.

When its usefulness has been realised (after the delay-line), this phase alternation has to be cancelled out by an inverting switch working at line rate. This is the PAL switch, and to keep it in synchronisation a further signal, called *ident* (short

Figure 19.6 *Position of the reference burst in the line waveform*

267

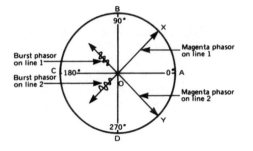

Figure 19.7 *Colour and burst axes for PAL encoding. The burst and V chroma signals are 'mirrored' about the axis AC on a line-by-line axis*

for identification), is sent. This is achieved by advancing and retarding the phase of the burst signal itself by 45° per line – the *swinging burst*. The burst phase alternations, then, are used in the TV's decoder to identify the PAL line (the one on which the R − Y carrier is inverted), while the reference oscillator locks to the *average* burst phase. Figure 19.7 indicates the relationship between burst phase and R − Y subcarrier phase.

Chrominance bandwidth

In transmission, the bandwidth of each colour-difference signal is limited to 1.2 MHz; this results in a rough, low-definition colour image which, however, is very acceptable when overlaid on the detailed luminance picture. This is because the human eye cannot perceive fine detail in colour, and the brain is fooled into believing that a full-definition, full-colour picture is present. We shall see that further liberties are taken with chrominance bandwidth in domestic VTRs, but still with subjectively acceptable results.

Colour signals in the VTR

If we had a perfect video tape recorder there would be no reason why we could not take the composite video signal, chrominance and all, and record it on the tape as an FM signal, demodulate it during replay and recreate the original video signal as described for luminance in Chapter 18. As we've already seen, a home VTR is far from perfect in its performance, and the direct recording idea will not work for two good reasons.

The bandwidth limitations of our VTRs have already been described, and with an upper cut-off frequency around 3.5 MHz, the colour signals at 4.43 MHz are 'left out in the cold' somewhat! Even in S-VHS format, whose signal passband approaches 5 MHz, it is still not possible to directly record an encoded colour signal. Because the VTR relies on mechanical transport, perfectly smooth motion cannot be achieved. Inevitable imperfections in bearing surfaces and friction drives, the elasticity of the tape and minute changes in tape/head contact all combine to impart a degree of timing jitter to the replayed signal. In good audio tape systems this jitter is imperceptible, and for luminance replay its effects can be overcome by the use of timebase correctors in professional VTRs, or fast-acting TV line-synchronising systems with domestic VTRs. The latter class of machine, in good condition, may be expected to have a timing jitter of about 20 μs over one 20 ms field period. Let's examine the effect of this on the chrominance signal.

We have seen that the hue of a reproduced colour picture depends entirely on the phase of the subcarrier signal. 'Phase' really means relative timing, so if the timing of the subcarrier signal relative to the burst is upset, hue errors will appear. One cycle of subcarrier occupies 226 ns, and a timing error of this order will take the vector right round the 'clock' of Figure 19.5, passing through every other hue on the way! A timing error of 100 ns (one ten-millionth of a second) will turn a blue sky into a lime-green one, and the jitter present in a mechanical reproduction system will make nonsense of the colours in the picture. For acceptable results, subcarrier phase errors need to be held within 5° or so, representing a permissible maximum timing jitter of ±3ns as shown in Figure 19.8. The physical causes of jitter cannot be eliminated so an electronic method of subcarrier phase correction is necessary.

Figure 19.8 *Hue is dependent on phase angle, and only a very small phase change is permissible in a VTR system*

Colour under

Let's first examine the problem posed by the high colour subcarrier frequency. Several solutions are possible, including decoding the chrominance signal to baseband (i.e. primary colour or colour-difference signals) before committing them to tape, then recoding them to PAL standard during the replay process. This would require a lot of circuitry, and be vulnerable to hue and saturation errors. A better solution is to keep the PAL-encoded signal intact, bursts, phase modulation and all, and merely convert it to a suitable low frequency for recording. During playback the composite chrominance signal can be re-converted to the normal 4.43 MHz carrier for recombination with the luminance component into the standard form of colour video signal, compatible with an ordinary TV set.

Space has to be found for the colour signal in the restricted tape frequency spectrum, however, and it is made available below the luminance sideband, where a gap is purposely left in the 0–1 MHz region. As explained in Chapter 18, this area is kept clear by tailoring the luminance record passband and the FM modulator deviation to avoid luminance sidebands falling into it during colour recording. The colour signal, then, is down-converted and bandwidth-limited so that it occupies a frequency band of roughly 0–1 MHz. It is a double-sideband signal, so that the carrier frequency needs to be centred on about 500 kHz and the sidebands limited to 500 kHz or so. The spectrum of the colour-under signal is shown in Figure 19.9; compare this with Figure 19.4.

Figure 19.9 *Spectrum of luminance and chrominance signals on the tape. The luminance signal is now in FM form, and the chroma signal still at baseband, but with limited bandwidth*

Down-conversion

The process for down-converting the chrominance signal is shown in Figure 19.10. The double-sideband signal is first separated from luminance information in a high-pass filter, then its bandwidth is limited to around 1 MHz by a bandpass filter. This curtailment of the sidebands of the modulated signal has the effect of restricting the chroma signal detail to around 500 kHz, less than half that of the broadcast signal. The chroma signal now passes into a mixer where it beats with a stable locally-generated CW signal at around 5 MHz. This is the familiar hetero-dyne effect, and the mixer output contains components at two frequencies, those of the sum and difference of the two input signals. A low-pass filter selects the required colour-under signal, and rejects the spurious HF product.

The colour-under signal contains all the information that was present in the 4.43 MHz chroma signal except the finer colour detail; thus chroma amplitude and phase modulation are preserved, as well as the phase and ident character-istics of the burst, albeit based on a much lower subcarrier frequency. Figure

Figure 19.10 *The basic colour-under system. A heterodyne technique is used to convert the chroma signal to a new, low frequency*

19.11*a* shows an oscillogram of the chrominance component of a colour-bar signal in 4.43 MHz form as broadcast, while in Figure 19.11*b* appears the same signal after down-conversion to a base frequency of around 500 kHz. Looking at the coarse structure of the waveform at *b* it is hard to believe that all information is preserved intact, especially the crucial burst signal, which appears as little more than two cycles of oscillation – it is so, however!

Up to now we have been quoting subcarrier frequencies in round numbers. Format specifications quote the colour-under frequency very precisely, and taking VHS as an example, the local CW signal is at a frequency of 5.060571 MHz, which when beat against the 4.433619 MHz broadcast subcarrier provides a colour under centre frequency of 5.060571 – 4.433619 = 0.626952 MHz, or 626.952 kHz. Due to broadcast subcarrier modulation, sidebands will extend for roughly 500 kHz on each side of this carrier.

For reasons which will become clear later, the local CW signal at 5.06 MHz needs to be locked (i.e. have a fixed phase relationship) to the line syncs of the recorded signal and this is achieved by a PLL (phase-locked loop).

Chroma recording

The down-converted colour-signal based (for standard VHS format) on 627 kHz, and containing amplitude and phase modulation, is recorded directly on to the tape. Its amplitude is carefully controlled to achieve maximum modulation of the tape without non-linearity due to magnetic saturation. In this respect it is similar to audio tape recording, with the necessary HF bias being provided by the FM luminance record signal – the two are added in the recording amplifier, giving rise to the waveform of Figure 19.12, where can be seen the recording-head signal for a colour-bar signal. The chrominance signal is superimposed on the constant-level FM luminance carrier, and this composite waveform is fed to the heads for recording on the tape.

Chrominance replay

During playback the chrominance signal appears at the replay amplifier in the colour-under form described above, and with a waveform like that in Figure 19.12. Impressed on to it are the timing variations, or jitter, introduced by the mechanical record and replay transport system and these have to be eliminated. First, though, the chroma signal is separated from the luminance FM signal in a

(a)

(b)

Figure 19.11 *(a) Oscillogram of encoded colour-bars as broadcast, based on a frequency of 4.43 MHz. (b) Down-converted colour-bars, now at a basic frequency about one-eighth of that in (a). The 'ghost' burst to the left of the real one is a spurious effect of the oscilloscope from which the photograph was taken, and is not actually present in the colour-under waveform*

low-pass filter. Up-conversion to correct subcarrier frequency (4.43 MHz) must also be carried out, and we'll take this first.

The 627 kHz is heterodyned with a local 5.06 MHz CW signal to produce a beat frequency of 4.43 MHz, selected by a suitable bandpass filter. The idea is shown in Figure 19.13. Dropout compensation and crispening are not relevant to the relatively coarse and 'woolly' colour replay signal, and the recreated 4.43 MHz chroma signal is added to the luminance component and passed out of the machine.

Figure 19.12 *The composite colour bar signal (FM luminance plus down-converted chroma) as applied to the recording heads*

Figure 19.13 *Basic chroma replay process of up-conversion*

De-jittering

As we have seen, the up-conversion process during replay involves a heterodyne system. If we can ensure that the 'local oscillator' is modulated by just the same timing errors as the off-tape chroma signal, the jitter on the replayed waveform will be cancelled out. The heterodyne process works by producing a *difference frequency* between two inputs (in this case the local 5.06 MHz CW signal and the off-tape chroma signal at 627 kHz). If both inputs contain the same jitter component they will move together in frequency and phase, so that the difference between them will be constant. This 'constant difference' is in fact the 4.43 MHz output signal, which will thus be unaffected by timing errors. On replay, then, the 5.06 MHz CW signal is known as the *jittering reference*. Its derivation will now be described.

The luminance signal recorded on the tape contains line sync pulses, and when these are recovered from the luminance FM demodulator they will contain the same timing errors, or jitter, as the chrominance signal, having been recorded and played back under the same circumstances and at the same time. If we separate them from the luminance part, they can be used to generate a suitable jittering

273

Figure 19.14 *A jittering reference generator using off-tape line sync and a phase-lock loop*

reference. Figure 19.14 shows how. Here we have a stable crystal oscillator running at 4.435 MHz as one input to a mixer, and a 625 kHz signal as the other. The 625 kHz signal is generated by a VCO (voltage-controlled oscillator). Now 625 kHz is exactly 40 times line frequency, so if we divide the oscillator output by 40 in a *counter* we have a line-rate signal suitable for comparison with off-tape line sync in a phase detector. This is another example of a PLL and its effect is to impress the jitter signal on to the VCO output. This jittering 625 kHz signal when beat against the stable 4.43 MHz crystal output will produce a jittering 5.06 MHz reference as required.

Let's take an example to illustrate the action of the circuit. Assume that a timing error has momentarily increased the off-tape chroma carrier frequency by 40 Hz, bringing it up to 626.992 kHz. Because colour-under frequency is 40 times line rate, the off-tape line sync pulse frequency will simultaneously rise by 1 Hz to 15.626 kHz, and this will appear at point X on Figure 19.14. The PLL will push up the VCO frequency to rebalance its phase detector and achieve 15.626 kHz at its Y input, and VCO output will rise to 625.040 kHz. The sub-mixer 2 will beat this against the steady 4.435571 MHz crystal frequency to produce an output of 5.060611 MHz. This passes into the main mixer 1 where it beats with the off-tape chroma carrier at 626.992 kHz to produce a final output frequency of 5.060611–0.626992 = 4.433619 MHz as was present with no timing error. Although this example describes a *frequency* error, phase errors (timing errors of less than one down-converted subcarrier cycle) are dealt with in the same way.

It is now plain why the colour-under signal is phase-locked to off-air line sync during record – the sync and colour signals are being recorded simultaneously, so that jitter cancellation works for both record and playback processes, the key in each case being the closely accompanying line sync pulse. The same PLL and 40-times divider is used in record and playback. The block diagram of Figure 19.15 shows the basic processes in chrominance recording, playback and jitter-compensation in a domestic VTR.

There are several other methods of deriving a jittering reference, all involving an off-tape signal of one kind or another. In the *burst-lock* system the colour-burst signal itself is used, being gated out of the replay chrominance signal and used to slave a local oscillator. Another system uses a pilot tone, laid down on the tape

Figure 19.15 *Overall view of the colour-under system*

during record, to achieve the same end. All current home formats, however, use the line-sync reference system described above, in conjunction with a 'tighter' jittering reference derived from the off-tape chroma sub-carrier signal itself – more details are given later.

ACC and colour-killers

Colour TV sets incorporate an ACC (automatic colour control) system to prevent tuning and propagation errors upsetting the luminance/chrominance ratio. This is necessary in a VTR for the same reasons, so that during record the ACC circuit monitors the amplitude of the colour-burst signal (transmitted at a constant level) and adjusts the gain of the chrominance record amplifier to maintain it at a constant level; by these means the chrominance signal itself is held at correct amplitude. To prevent the recording of *confetti* or unlocked colours during monochrome transmissions or when the record down-conversion PLL is unlocked, a record colour-killer is provided, and this is sometimes linked to a rear-mounted switch so that killer action can be invoked manually for a full-bandwidth monochrome recording of any signal which contains colour information.

On replay, too, these functions are required. The off-tape chrominance signal is vulnerable to level fluctuations and these are ironed out by the ACC loop. The replay colour-killer (which again can be manually overridden in some machines) acts to prevent spurious colouring of replay pictures from monochrome videotapes, and shuts down the chrominance channel in the event of malfunction of the replay up-conversion processes.

275

Figure 19.16 *The effects of azimuth tilting at different frequencies. LF signals have low immunity to crosstalk*

Crosstalk compensation

In Chapter 17 we saw that early machines used a guard band between recorded tracks on the videotape to prevent the possibility of crosstalk between adjacent tracks. Later formats eliminate this guardband and achieve greater information density on the tape by laying down tracks with no intervening safety gap. To prevent crosstalk between tracks each head is given an azimuth offset; this ensures that readout from adjacent tracks is at a very low level and does not interfere with reproduction. As Figure 19.16 shows, however, the azimuth-offset idea is less effective at low frequencies, and the 0–1 MHz region occupied by the chroma signal will suffer unacceptable crosstalk between tracks unless steps are taken to remove the interfering signal.

We cannot prevent inter-track crosstalk taking place at colour-under frequencies, so methods of electronic cancellation of the unwanted signal have been devised. Each format uses the same basic idea but differs in the way that it is carried out. Let's look at a typical cancellation system, that used in the VHS format, to explain the principles involved.

The interlaced scanning system employed in broadcast TV calls for a half-line offset between the starting points of odd and even fields, as explained in Chapter 1. If the video tracks were laid down on the tape in this form, there would be a half-line offset between the recorded TV lines, just like that shown on the TV screen in Figure 17.14. However, it is not that simple, as the video tracks are laid on the tape at a slant angle, and the tape moves horizontally past the drum. The combined effects of these result in the individual lines of each TV field being laid exactly alongside each other, as shown in Figure 19.17. This ensures that when crosstalk occurs the interfering signal is coming from the 'same' line on the partnering field, and has the *same timing*, within that line, as the wanted signal. Thus there will be a fixed phase relationship between wanted and crosstalk signals, bearing in mind that the colour-under signal is based on a frequency (for the VHS system) of 626.9 kHz.

Figure 19.17 *The tape tracks are arranged so that individual TV lines lie physically adjacent on the tape*

During record, head A is fed direct by the colour-under signal, and the PAL characteristic of this signal will result in the burst phases being laid down on the tape according to (a) in the chart of Figure 19.18. Thus TV line *n* has its burst at 135°, line *n*+1 at 225°, line *n*+3 135° and so on. The chroma signal fed to head B is *delayed* by 90° on each successive TV line as shown in line (b) of the chart. Thus line *n* is 'normal', line *n*+1 has a 90° clockwise shift, line *n*+2 is rotated 180° clockwise, line *n*+3 270° clockwise, line *n*+4 normal, and so on throughout the field period. During replay, head A will pick up a large signal from its own track and a smaller crosstalk signal from the adjacent tracks laid down by head B, with burst phases as shown in line (c) of the chart. To recover a 'normal' signal from head B during replay, we have to *advance* the phase of its output by 90° per TV line, to compensate for the 90° retard per line introduced during recording. Thus head B's corrected output will be line (b) on the chart, but phase advanced by

Figure 19.18 *Burst phasor diagram for the VHS colour-under system. Lines (c) and (d) can be seen to emerge with correct phasors as broadcast*

277

Figure 19.19 *Delay line matrix for chroma crosstalk cancellation*

90° per TV line, as shown on chart line (d). The small arrows here show the crosstalk signals picked up from track A, *having undergone the same 90° per line phase advance.* Study of playback lines (c) and (d) of the chart will show that if we compare *any* playback TV line with its next-but-one neighbour, the main signal vectors add (i.e. they are pointing in the same direction), whereas the crosstalk vectors (small arrows) cancel. This is illustrated by the hatched boxes for head A playback (TV lines *n* and *n*+2) and the dotted boxes for head B playback (TV lines *n*+5 and *n*+7).

The reason for the carefully-contrived head B recording sequence is now clear. To eliminate the crosstalk signal it is only necessary to *add* each TV line to its next-but-one neighbour, which can be brought into time coincidence by a two-line (128 µs) delay, then added in a suitable matrix, as illustrated in Figure 19.19. The process is very similar to that which takes place in the familiar delay line circuit of a colour TV receiver, and the glass delay line is similar in construction (but with twice the path-length) to its TV counterpart.

We have described the crosstalk-compensation process in terms of burst phasors, whereas our real concern is eliminating crosstalk between the chrominance *signals* on adjacent tracks. Since the chroma signal is defined (in its encoded form) in amplitude and phase, the crosstalk compensation system works equally well on the subcarrier signal as on the burst itself.

Before we go on to describe the electronics in the chroma phase-shifting process, we must recap the relevant parameters of the VHS format. The colour-under frequency is 626.952 kHz, produced in a balanced modulator (mixer) by beating the incoming 4.433619 MHz colour signal with a locally generated 5.060571 MHz CW signal containing the necessary phase-shifting information for recording head B.

Figure 19.20 shows the system in block diagram form. This is similar to the colour-under system we met earlier in this chapter, with a main mixer 1 producing the colour-under signal and sub-mixer 2 acting to produce the 'local' signal.

For VHS, mixer 2 is fed from the output of a stable free-running crystal oscillator at 4.435571 MHz; and a 625 kHz signal (40 × line frequency) generated by a VCO slaved to incoming line sync. On its way to mixer 2, the 625 kHz signal passes through a phase shifter (part of a purpose-designed IC) which *advances* its phase by 90° per line for one field (that recorded by head B) but is reset to zero phase shift for the next field (head A's province). The 'reset' signal comes from a tachogenerator pulse which is derived from a magnet on the head drum, and which indicates that head A is writing. This 90° line-by-line phase advance, operating on each B track only, passes intact through additive sub-mixer 2, and

Figure 19.20 *The VHS colour recording system to give the tape track pattern of Figure 19.18. Operation is explained in the text*

is turned into a phase *retard* at subtractive main mixer 1 for application to recording head B.

To sum up, then, we have going on to the tape a down-converted subcarrier signal based on 626.9 kHz, with sidebands extending down to about 100 kHz and up to about 1.1 MHz. This subcarrier is identical to the encoded broadcast colour signal, but based on the new, lower frequency. Head A records odd-numbered tape tracks without modification, whereas head B records the subcarrier signal on even-numbered tape tracks with a phase retard of 90° on each TV line. The recording signal for head B is modified in this way to facilitate crosstalk elimination in a two-line delay matrix during replay.

VHS chroma replay

Figure 19.21 shows a simplified block diagram of the replay process. The off-tape 626.9 kHz chroma replay signal enters at the top left, and is mixed in main mixer 1 with the local 5.06 MHz signal to produce a difference signal of 4.43 MHz, selected by the bandpass filter. This forms the up-converted output signal. The local 5.06 MHz signal needs to have a phase-stepping characteristic to compen-

279

Figure 19.21 *VHS replay system block diagram showing up-conversion, line-rate de-jittering and phase restoration*

sate for that introduced during record. Once again, sub-mixer 2 is used, and operates, as on record, to beat two locally produced signals – these being the 4.435571 MHz crystal oscillator output; and the 625 kHz (40 × line frequency) signal from the PLL. The action of the phase shifter is the same as on record (phase advance of 90° per line for head B only) but because both mixers 1 and 2 are now operating in *subtractive* mode the phase advance effect emerges unchanged and appears as a phase retard effect imparted during the record process.

The recovered and up-converted chroma signal now passes into a two-line (128 µs) delay line matrix, where crosstalk cancellation takes place to render a 'clean' chrominance signal at 4.43 MHz, ready to be added to the luminance signal.

VHS de-jittering

The off-tape chroma signal at 626.9 kHz contains a timing error (jitter) component, and this has to be removed during replay. As in the 'basic' process described earlier in this chapter, this is achieved by using a jittering reference from off-tape line sync. The reference signal for the 625 kHz PLL at the bottom of Figure 19.21, then, comes from the replay luminance signal via a sync separator, and the 625 kHz input to sub-mixer 2 is thus slaved to off-tape sync, and contains a jitter reference as well as the 90° per line phase stepping effect for head B. The jittering reference passes through sub-mixer 2 and cancels the chroma jitter in main mixer 1.

A further de-jittering process is carried out in VHS machines, and this is shown in Figure 19.22, an enlargement of the previous diagram Figure 19.21. Here we

Figure 19.22 *Expansion of Figure 19.21 to show the second de-jittering loop, which works at subcarrier rate. Newly introduced blocks are shown in bold outline*

see the additional circuit blocks on the right, and their operation will now be explained. A sample of the 4.43 MHz chroma output is taken and passed into a burst gate which extracts the colour burst signal. This burst is compared with the output of a stable 4.43 MHz crystal oscillator in phase detector 1, any jitter present giving rise to an error signal which 'steers' the frequency of the 4.435571 MHz crystal oscillator (now working as a voltage-controlled oscillator, VCO) in a direction to compensate for the jitter, and cancel it out. If the phase error becomes too great for phase detector 1 to cope with, phase detector 2 comes into play, sending a pulse to the phase-select circuit to shift the 625 kHz signal by 180°, thus pulling it back within the range of phase detector 1. The net result is that the 4.43 MHz chrominance output signal is slaved to the stable 4.43 MHz crystal oscillator reference.

A colour-killer output is taken from phase detector 2, and this ensures that replay chrominance is only enabled when the VTR's chrominance output is correctly locked to the local 4.43 MHz crystal oscillator.

Video-8 format chrominance

The basic principle of the colour system of Video-8 format is the same as for VHS already described, in that the chroma phase recorded on tape is manipulated to ensure that crosstalk signals come off the tape in antiphase over a two-line period. The colour-under frequency for V8 is $(47 - \frac{1}{8})\, f_h$, which is 732 kHz. It is locally generated, but in order to implement de-jittering during playback, is locked to incoming line sync in a PLL incorporating a $\div 375$ stage. We start, then, with $f_h \times 375 = 5.86$ MHz. This is divided by eight in a counter to render 732 kHz; the counter is under the influence of drum flip-flop and line-rate input pulses so that the phase of the colour-under carrier is advanced by 90° per line for head A's

281

	n	n + 1	n + 2	n + 3	n + 4	n + 5	n + 6	
1	⇧	⇦	⇩	⇨	⇧	⇦	⇩	Head A record: phase advances 90°/line
2	⬆	⬆	⬆	⬆	⬆	⬆	⬆	Head B record: constant phase
3								Head A playback: contains crosstalk from B
4								Head A playback: correct phase restored
5								Head A playback: delayed by 2 lines
6								'A' output from delay-line matrix: crosstalk removed

Figure 19.23 *The stages in removing crosstalk interference from the Video-8 chroma signal. The 'n' numbers refer to television scan lines*

sweeps only; see row 1 of Figure 19.23. Head B's chroma record signal is not phase-changed, as row 2 of Figure 19.23 shows.

During replay the chroma signals are read-out from tape according to row 3 in Figure 19.23: this shows the output from the Ch1 (a) head, and represents the main signal as recorded, together with a crosstalk contribution from adjacent B tracks, shown as small black arrows. To cancel the effect of the phase advance given to head A's chroma signal during record a corresponding phase-retard of 90° per line must be imparted to it during playback. This has the effect of restoring phase normality to the chroma *signal*, but introducing a 'twist' to the crosstalk component as shown in row 4 of the diagram. The process has not removed the crosstalk, but has paved the way for its cancellation.

Row 5 of Figure 19.23 shows the new timing of row 4 after its passage through a two-line delay system. Adding the delayed and non-delayed signals, so long as their timing is exactly right, gives addition of the wanted (in-phase) chroma signals, but complete cancellation of the crosstalk signals which are now in opposite phase for every line. The resultant 'clean' chroma signal is shown in the bottom row. Although we have only illustrated here the situation for a pure red signal recorded by head A, the same crosstalk-cancellation mechanism works for any colour, and also for head B's signals from which crosstalk from A tracks is removed.

The only time the system breaks down is when one line is markedly different in hue from its next-but-one neighbour. Under these circumstances of colour changes at horizontal edges in the picture the crosstalk-compensation system will introduce hue errors centred on the point of transition, the effect being worse than complete loss of crosstalk cancellation. To prevent this the chroma two-line delay circuit is governed by a *correlation detector* which looks for large disparities in hue over a two-line period. When they are detected the effect of the chroma delay line is cancelled.

Advanced techniques are applied elsewhere in the Video-8 colour circuitry. The burst signal, a crucial reference for both VCR and TV, must be kept in the best possible condition. To give it better immunity from tape-noise its amplitude is doubled during record, and restored to normal during playback. In addition to this *burst-emphasis*, the chroma signal itself is treated for noise-reduction in a fre-quency-conscious amplifier which gives increasing emphasis to the outer 'skirts' of the chrominance sidebands. Corresponding de-emphasis during playback gives a useful overall reduction in 'confetti' on the monitor screen. Indeed, these record techniques ensure that most chroma signals on tape, regardless of their amplitude in the 'real' signal, are recorded at high levels, clear of the 'noise floor.'

Chrominance definition of replayed pictures

We saw at the beginning of this chapter that the bandwidth of the chrominance signal is restricted during its passage through the recording system of a domestic VTR, resulting in a reduction of horizontal resolution (the ability to define vertical coloured edges) by about 50 per cent. The effect of the delay line matrices in the chroma crosstalk compensation system is to impair the vertical chrominance resolution, and the ability of the system to define horizontal coloured edges (vertical definition) is in fact reduced by a factor of eight overall. This has the effect of rendering roughly equal the vertical and horizontal chrominance defini-tion of the displayed picture on replay, and is considered acceptable. There is little point in maintaining good vertical resolution when horizontal definition is restricted, though this is done in the broadcast signal, where vertical resolution is governed by the number of TV lines in the picture, whereas horizontal resolution depends on chroma channel bandwidth – 1.2 MHz or thereabouts in the UK system.

PAL encoding and crosstalk compensation

Although we have used the PAL chroma and burst vectors to illustrate the prin-ciple of the chrominance crosstalk compensation system, the PAL characteristic of the encoded chroma signal takes no part in the crosstalk compensation process. Thus the cancellation systems described above work equally well on any form of colour-encoding system. Because the PAL signal (and other line-alternating colour-encoding systems such as SECAM) has a two-line pattern, the crosstalk compensation is carried out on a two-line basis to retain the swinging burst (and chroma V) characteristics. With a simple encoding system such as NTSC, a one-line crosstalk cancellation system would be possible, conferring the advantage of better vertical chrominance resolution.

20 *Signal processing: audio*

For several years the audio system used in domestic VTR formats depended on a longitudinal track laid along the extreme top edge of the ribbon. Figure 17.5 shows its position, and Figure 17.9 the placement of the stationary head which writes and reads the audio track. One of the virtues of home VTRs is their low tape consumption, but this depends on a slow tape speed, much slower for instance than an audio compact cassette. The result is a constriction on audio frequency response, particularly in LP modes where the rate of progress of the tape is halved.

The noise performance, too, of a longitudinal sound recording on video tape falls far short of other sound systems. This stems partly from the low tape speed (massive equalisation must be applied to the off-tape signal to maintain flat frequency response) but mainly from the narrowness of the tracks necessarily used. The standard longitudinal track width is 1 mm, reduced to two of 0.35 mm in the ephemeral stereo versions of VTRs using this system. Figure 20.10 highlights the shortcomings of VTR longitudinal track performance, and compares them with the results obtained from the new systems to be described now.

FM audio

The ingredients for success in recording TV pictures on tape are high writing speed and FM modulation as we saw in Chapters 15 and 16. Within the immovable constraints of existing formats it was obvious that for better sound performance these virtues must also be applied to the audio signal. The obvious solution was to record an FM-modulated sound carrier in or around the helical video tape tracks; but with the vision heads and tracks already chock-full of information (see Figures 17.30 and 19.9) the problem was where to squeeze in the audio recording with regard to frequency-spectrum space and the 'magnetic' capacity of the tape track. Some form of multiplex system is required. We have already met this in the description of MAC TV systems in Chapter 10, where a method of getting several streams of information through a single channel path was found in time-compression and time-division-multiplex. A similar technique is in fact used in the PCM sound systems to be discussed later in this chapter. All TDM setups require signal storage at each end of the link, however. Two alternative systems emerged for home VTR formats – *depth multiplex,* used in VHS models; and *frequency multiplex* in Video-8 format. The former is the most common, and will be covered first.

Depth-multiplex audio

This technique depends to some degree on the magnetic layer of the tape itself to discriminate between the video and audio signals. A separate pair of heads on the spinning drum is provided solely for the audio signal FM carriers. A typical layout of heads on the drum is shown in Figure 20.1. The two audio heads are mounted 180° apart and are arranged to 'lead' the video heads. The gaps cut in the audio heads have large azimuth angles: ±30°, sufficient to prevent crosstalk from adjacent tracks at the carrier frequencies (around 1–2 MHz) involved. The mounting height of the audio heads is set to place the hi-fi audio track in the centre-line of the corresponding video track. This track layout is shown in Figure 20.2, which also gives an idea of the relative audio track widths; the audio tracks are half the width of the vision tracks they accompany.

Figure 20.3 gives an idea of the depth-multiplex principle. Here the heads are moving towards the right across the tape, led by the audio head which has a relatively wide gap. The effect of this is to write into the tape a deep magnetic pattern penetrating some microns into its magnetic coating. Shortly following the audio head comes the regular video head, whose gap is in the region of 0.25 micron. Writing (for the most part) at higher frequencies, it creates shallower,

Figure 20.1 *Depth-multiplex recording drum. This one has the audio heads mounted 138° ahead of the double-gapped video heads*

Hi Fi sound
track 26μ
−30°

Video
track
49μ +6°

(a)

Figure 20.2 *Video/sound track relationship for VHS*

Video signal

Audio signal

Surface layer:
recorded video

Deep layer:
recorded audio

Tape magnetic medium

Tape base

Figure 20.3 *Depth-multiplex principle: separate magnetic layers are written into the tape*

shorter magnetic patterns which penetrate less than 1 micron into the magnetic surface. Thus the recorded tape (top LHS of Figure 20.3) contains a 'two-storey' signal: a buried layer of long-wavelength audio patterns under a shallow top layer of video patterns.

During replay the same heads operate on the same tracks as before. The video head picks off its track with little impairment, and with minimum crosstalk from sound tracks because of the large disparity (36° for VHS-SP) between the azimuth angle of recorded track and replay head. The audio head during replay is handicapped by the barrier presented by the 'video layer' on tape, but the resulting 12 dB or so of attenuation – thanks to the use of bandpass filters and the noise-immunity of the FM carrier system – does not prevent noise-free reproduction of the baseband audio signal as long as the tracking is reasonably correct. For hi-fi sound the tracking performance is critical, especially in LP modes where narrow tracks tend to magnify any errors.

As with the video heads, switchover between the two rotating audio heads is carried out during the overlap period when both heads are momentarily in contact with the tape, one just leaving the wrap and one just entering. Because of the angular offset between video and audio heads on the drum (Figure 20.1) the head switchover point for the latter is set by a second, *delayed* head flip-flop square-wave triggered from the drum tacho-pulse. The FM carrier signals to and from the audio heads on the drum require their own rotary transformers, which may be concentric with those for the video signals under the drum; or mounted above as in the photo of Figure 20.4.

286

Figure 20.4 *Head drum mounting in a Panasonic hi-fi VHS machine. The rotary transformer for audio signals is above the drum*

The frequencies used for VHS stereo hi-fi in depth-multiplex systems are 1.4 MHz (L) and 1.8 MHz (R); the FM modulation sidebands extend for about 250 kHz on each side of the (unsuppressed) carriers. Each audio head deals with *both* FM carriers throughout its 20 ms sweep of the tape – during replay the carriers are separately intercepted by bandpass filters for processing in their own playback channels.

Frequency-multiplex audio

An alternative approach, used in Video-8 VTRs, to helical sound recording is to use the *video heads* to lay down on the tape an FM audio soundtrack.

The baseband frequency response is limited to about 15 kHz. As Figure 17.30 shows the audio FM carrier is based on 1.5 MHz and has a deviation-plus-side-band width of about 300 kHz. It is possible to insert this 'packet' between the outer skirts of the upper chrominance and lower luminance sidebands, permitting it to effectively become part of the video signal so far as the heads and tape are concerned. The FM audio signal carrier is added to the FM luminance signal in the recording amplifier, and laid on tape at a low level – some 18 dB below that of the luminance carrier. This suppression of sound carrier level helps prevent mutual interference between sound and vision channels, whose outer sidebands overlap to some degree.

The original AFM recording system made provision only for monaural sound. The advent of hi-fi stereo recording in the competing VHS format led to the adoption of a stereo AFM plan for Video-8 formats, using a second sound carrier at 1.7 MHz. Unlike VHS hi-fi, however, it is not possible here to simply use one carrier for each channel because compatibility with mono AFM equipment must be maintained. The stereo-AFM system is illustrated in Figure 20.5. Incoming L

Figure 20.5 *Record and playback signal processing for AFM stereo audio in Video-8 format*

and R audio signals are brought together in an adder, and the result halved in an attenuator to derive an (L+R)/2 signal for FM modulation and recording on tape at 1.5 MHz. This forms the compatible mono signal. The L and R record signals are also routed to a matrix in which they are *subtracted* to produce an L − R signal, which is now halved to render an (L − R)/2 signal. It is this which is processed and frequency-modulated onto the 1.7 MHz carrier. During replay the recovered L + R and L – R signals are added and subtracted in separate matrices to derive L and R signals. Readers who are familiar with VHF-FM analogue stereo sound broadcast/ reception techniques will recognise this 'stereo-difference' technique – in radio broadcasts the difference signal is conveyed in a subcarrier at +38 KHz.

Playback carrier routing

In replay mode the separation of the various off-tape signals in Figure 17.30 is carried out by four separate filters: a low-pass type with cutoff around 180 kHz for ATF tones; a low-bandpass filter centred on 732 kHz for colour-under signals; a narrow bandpass one tuned to 1.5 MHz for interception of the audio-FM carrier; and finally a high-pass (roll-on about 1.7 MHz) acceptor for the FM vision signal with its sidebands.

Audio electronics

In 'longitudinal' audio systems the only processing necessary for the audio signal itself is some recording equalisation and the addition of a bias signal as described in Chapter 16. For hi-fi recording a great deal more work must be done on the input signal to condition it for multiplex recording, and to reduce the overall system noise level to a point far lower than can be achieved with conventional sound recording systems on tape or disc. The techniques used are almost identical for depth-multiplex and frequency-multiplex systems. In both cases much of the audio processing circuitry is common to both record and replay.

Figure 20.6 shows, in generalised form, the arrangements for a single channel of hi-fi sound processing. In record the audio signal (left of diagram) is first passed through a high-cut filter to remove supersonic frequencies, then via switch S1 to the non-inverting input of an operational amplifier. The op-amp output (ignoring for a moment S3) takes two paths: one through a pre-emphasis 2 network to the input of a VCA (voltage controlled attenuator); and one via a weighting filter to precision RMS-level detector. The output of this device reflects, from moment to moment, the true RMS (root-mean-square) amplitude value of the signal it sees. The RMS detector output is now applied as a control potential to the VCA. Thus the output from the VCA is large for high-level audio signals, small for low level signals. It is routed via S2 to the inverting input of the op-amp. The effect of this variable negative feedback is to 'compress' the dynamic range of the audio signal emerging at the op-amp output. The compression ratio is 2:1, expressed in deci- bels on the left-hand side of the diagram in Figure 20.7. The amplitude- compressed signal, since it will be conveyed by an FM carrier, has carefully- tailored pre-emphasis characteristic imparted by the emphasis-1, emphasis-2 and weighting filter sections in Figure 20.6.

Continuing with that diagram, the compressed and pre-emphasised audio signal now passes through S3 to a limiter whose purpose is to clip any signal excursions that may cause over-deviation in the FM modulator. Next follows an attenuator for pre-setting, via S4, of FM deviation in the VCO which performs frequency mod-

Figure 20.6 *Block diagram of hi-fi audio signal processing, fully explained in the text*

Figure 20.7 *Overall effect of companding on the audio signal. The dynamic range is greatly compressed for the 'on-tape' phase*

ulation. The FM carrier is conveyed by S5 to a low-pass filter whose output is added to ATF, chrominance and luminance-FM recording signals for passage to the video heads (V8 format); or passed directly to the rotary audio heads in VHS-format machines.

As is obvious from Figure 20.6 all the important processing blocks are used again during playback. Replay head input is now routed by S5 to a bandpass filter, centred on 1.5 MHz for Video-8, 1.4 MHz for VHS L-channel, 1.8 MHz for VHS R-channel etc. With its loop completed by S4 the VCO now forms part of a PLL detector whose demodulated output signal goes via S6 and S1 to the non-inverting input of the main op-amp, the gain of which is now set by the fixed feedback resistor from S3 to the *inverting* input.

Again the op-amp output signal is passed via the emphasis 2 filter to the VCA; again the latter is under the control, gain-wise, of the RMS detector whose input is conditioned in turn by the weighting filter. Once more the output of the VCA passes through the emphasis-1 circuit, now to form the audio output signal via S2. This time, however, the entire circuit within the dotted outline in Figure 20.6 is in the direct signal path rather than part of a negative feedback loop as it was during record. As a result its operation is reversed to 'mirror' its previous functions. Thus purely by the action of S2 what was the compressor/pre-emphasiser has become an expander/de-emphasiser, using all the same filters and components. Hence the expression *compander* and the symmetrical appearance of the diagram in Figure 20.7 whose right-hand side shows the effect of expansion of the signal's dynamic range during replay.

The most significant benefit of the companding circuit may not be obvious as yet. Plainly, it is advantageous to boost the low-level components of the audio signal so that they *significantly* deviate the FM carrier to swamp noise. The 'tape-noise' in an FM system is typically –45 dB, but since the lowest-level signals coming off tape are effectively 'attenuated' during playback, this –45 dB noise floor is depressed to –80 dB in practical systems. This ingenious combination of FM and companding techniques permits us to achieve an overall dynamic range and S/N ratio better than that inherent in the tape system itself as defined in Chapter 16 and Figure 16.7.

Dropouts can seriously upset the operation of a hi-fi audio system, and masking of these is even more important than for video. In Figure 20.6 a dropout detector (top right) monitors the off-tape audio FM carrier and throws switch S6 whenever it sees one, whereupon the 'hold 2' circuit and associated low-pass filter provide dropout compensation. Hold 1 circuit is primarily concerned with masking the dropout and disturbance associated with head-switching. If the dropout count becomes too high (e.g. severe mistracking) a mute circuit comes into operation; the audio output now reverts to the longitudinal sound track which is always provided as back-up, and to confer compatibility with other machines and tapes.

PCM audio

A brief outline of the method of quantising analogue signals was given in Chapter 11. All newly-developed hi-fi audio systems use digital encoding: examples are compact disc, the Nicam TV stereo plan, DAB radio broadcasting and MAC TV sound transmissions. Although digital transmission and recording requires more bandwidth than other systems it has the advantages that the two-state signal is more robust than its analogue counterpart, and (so long as the quantising rate is high) is capable of superb S/N ratio and wider dynamic range.

The Video-8 format has provision for stereo PCM in addition to the monaural AFM sound facility described above. The most expensive V-8 camcorders and homebase VTRs are fitted up for PCM operation. It is impractical to give full details of the system in this book, and what follows is a basic outline of the technique used.

Figure 20.8 shows the functional blocks in the PCM sound processing. Initially the audio signal for record is amplitude-compressed in a compander similar to that in Figure 20.6. Next follows a quantisation process in which it is sampled at $2 f_h$, (31.25 kHz) to 10-bit resolution. Since the tape system here cannot cope with 10-bit data a conversion is carried out to 8-bit: the process is a non-linear one, in which low-level signals are in effect given 10-bit (1024-level) descriptions, falling in three stages – as signal level increases – to 7-bit data (128-level) for the largest signal excursions. This has the effect (during playback) of concealing quantising noise in the loudest sound peaks, where they go virtually unnoticed; indeed the overall S/N ratio is subjectively equivalent to 90 dB.

Unless their transmission/recording media are very secure, digital data systems need error-correcting artifices to repair or conceal corruption of the data by noise and distortion. About 38.5 per cent *redundancy* is imparted to the PCM audio data by the addition to it of further data in the form of a cyclic redundancy check code (CRCC) – used during playback for corruption-test and 'first-aid' purposes. The data rate is high, and the effect of a tape dropout would ordinarily blow a hole in the information stream; to prevent this the data is 'scattered' on tape according to a cross-interleave code (CIC), part of the Video-8 format. The effect of a tape dropout thus becomes distributed during replay, and the 'frayed edges' can be repaired by use of the CRCC and an additional parity check system.

The 8-bit data words are now temporarily stored in a pair of 16K RAM memories. Writing to memory is performed in real time. Readout from memory is much

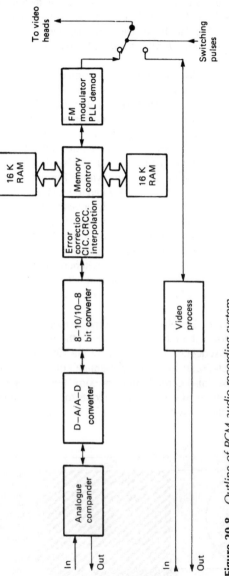

Figure 20.8 Outline of PCM audio recording system

faster: all the data (which contains information on both stereo channels) is clocked out in less than 3 ms at 20 ms intervals. The effect of this 7:1 time-compression is to push up the data-rate to about 2 Mbits/second, but to confine the audio transmission period to a small time-slot, in very similar fashion to that of the MAC system in Figure 10.5. As we shall see in a moment, a separate place is found on the videotape for this data-burst. The two-state data signal is tone-modulated at 2.9 MHz for data 0 and 5.8 MHz for data 1; this is called frequency shift keying (FSK).

At this point the PCM signal is ready to go onto tape. The switches on the right of Figure 20.8 change over once per field period, in synchronism with TV field rate and RAM readout. This feeds data to the video heads alternately, during a period when each is scanning a 'forward extension' of the helical vision track on tape – this is illustrated on the right of Figure 20.9. The conventional video tracks (which also contain AFM audio information, duplicating the PCM sound track) are recorded over 180° of the head track, but the head/tape angle is such that they occupy about 5.4mm of the 8mm tape width. The extra 30° or so of tape wrap shown on the right of Figure 20.9 is devoted to PCM recording: while one head is writing PCM data, the other (diametrically opposite on the drum) is recording the last lines of the TV picture at the top of the *video* tracks on the right of the diagram.

During playback, head switching ensures that PCM data read off the tape is routed to the audio section for the appropriate 30° scan/3 ms. As in Figure 20.6 the record VCO FM modulator is now switched to perform PLL demodulation, producing binary data at its output. Continuing to the left in Figure 20.8, the data, still in time-compressed form, is read into the same pair of 16 K memories as were used during record.

Memory readout takes place in 'real time', expanding the data to give *continuous* data at a lower bit-rate – that at which the memories were loaded during record. Readout sequence is governed by the CIC (see above) in order to de-interleave the data, scattering and fragmenting errors in the process. In the memory control chip is also carried out error correction by means of the CRCC and parity checks mentioned earlier. For dropouts too severe to be repaired by these means, *interpolation* provides a 'patching' system which, in the face of sustained

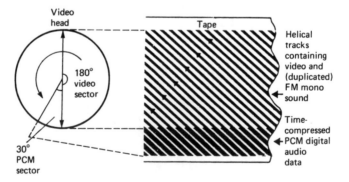

Figure 20.9 *The PCM data is recorded on a 'forward-extension' of the video tape tracks; an extra 30° of head rotation is reserved for this*

and continued corruption, devolves to a PCM mute action, switching the audio output line back to the AFM sound track. If this is also corrupted, silence will ensue!

The 8-bit 'reconditioned' data now passes to the 10-bit conversion stage to make ready for D/A conversion. As is common in these designs the record A/D converter is used for this, now switched to perform D/A operation. The analogue signal reconstituted at the D/A converter output is still in amplitude-compressed form, and is now expanded to full dynamic range in a logarithmic compander – the dotted-box section of Figure 20.6.

The relative performance of the audio systems described in this chapter is charted in Figure 20.10, where (a) shows the frequency responses, and (b) the S/N ratio capabilities.

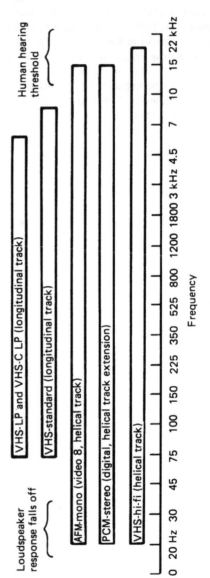

(a)

Figure 20.10 *Audio system performance of various formats. Bar-chart (a) illustratres frequency response capability; bar-chart (b) shows signal to noise ratio. The upper two bars in (b) are shown in 'raw' state, i.e. without the sophisticated noise-reduction systems usually applied*

Stereo VHS standard (longitudinal tracks)

Mono VHS standard (longitudinal track)

Stereo VHS-hi-fi (helical track)

AFM mono (Video 8) (helical track)

Stereo PCM (Video 8) helical extension)

| 15 dB | 17 | 20 | 25 | 32 | 40 | 50 | 65 | 80 | 100 dB |
| | | S/N ratio 10:1 | | 40:1 | 100:1 | | 1000:1 | | 10,000:1 100,000: |

(b)

21 *Servo systems and motor drive*

In an audio tape recorder, we have a single drive system to rotate the capstan, and this pulls the recording tape past the sound head at a constant, fixed speed. Provided this speed is the same during record and playback, the programme will be correctly reproduced, and if the tape speed is arranged to conform to the standard (4.75 cm/second for audio cassette machines) we have the further advantage of interchangeability between tapes recorded and played back on different machines. The only requirements, then, of a drive system in an audio tape recorder are that it should run at a specific speed, and that short-term speed variations (wow and flutter) are kept at a low enough level to prevent noticeable changes of pitch in the reproduced sound. This simple drive system is possible because an audio waveform carries *all* the necessary information about the sound signal at any one instant, and this waveform is laid down along the tape as a single longitudinal stream of magnetic patterns.

The situation with TV signals is rather different. Because the TV image is a two-dimensional display, and only a single transmission channel is available, the image has to be built up by line-by-line scanning of the scene, with $312\frac{1}{2}$ lines making up one field, and taking 20 ms to do so, and two fields interlacing to form a complete frame, or picture. Thus for television signals it takes 40 ms to build up one picture, and we have introduced a second factor into the signal: time. The video signal, then, has built-in markers to indicate this second dimension of time, and the markers which signify the beginning of each new field every 20 ms are very important in video tape recording, as we shall see.

Another factor in the process of recording TV signals, which only has an audio-recording counterpart in DAT machines, is the idea of laying down parallel slanting tracks on the tape by means of a rotating head drum. If replay of these tracks is carried out in a random manner (even at the correct speed) there is no reason why the replay heads should follow the path taken by the heads during record, so that each replay head may well read between the lines, as it were, and reproduce a signal consisting of a mixture of the information from two adjacent tracks. We have already seen that adjacent tracks contain different video signals recorded at different azimuth angles, so that this form of mistracking will make a noisy mess of the reproduced signals.

To overcome these problems, a very precise control system, known as a servo, is required to govern the head drum and capstan drive system. In this chapter we shall examine the purpose and mode of operation for both head drum and capstan servos, and see how they govern the relationship between video head position and tape track placement.

Requirements during record

We have seen in Chapter 17 that two video heads are used in home VTRs, each of which lays down one complete field of $312\frac{1}{2}$ TV lines during its scan of the tape. At some point during the field period one video head has to 'hand over' to the other, and as this will cause a disturbance on the reproduced picture, the 'head interchange' or switching point is arranged to lie at the extreme bottom of the picture, just before the field sync pulse (Figure 17.17). This means that the disturbance will be just off the screen in a correctly set-up monitor or TV set due to the normal slight vertical overscan. This means that during record we need to arrange for each head to start its scan of the tape about half a millisecond (corresponding to approximately eight TV lines) before the field sync pulse occurs in the signal being recorded; and hold this timing, or phase relationship, steady throughout the programme.

If this relationship were not present, the head-switching disturbance effect would move into the active picture area, and drift up or down the screen. The mechanics of the deck arrangement and the position of the 'entry guide' determine the point at which the moving tape starts its wrap of the head drum, as shown in Chapter 17. The electronic servo loop sees to it that each head passes this point and starts to write its field of information just before the field sync pulse occurs in the video waveform being recorded. To achieve this, we need some marker to tell the servo system the angular position of the head drum. There are many possible ways of generating this, but the most common is the use of a small permanent magnet fixed to the underside of the head drum, and arranged to induce a pulse in a nearby stationary pickup coil at each pass of the magnet, as shown in Figure 21.1.

Other methods of generating these pulses are possible. In some VTRs a drum-mounted 'flag' interrupts the light-path of an *optocoupler* once per revolution; or

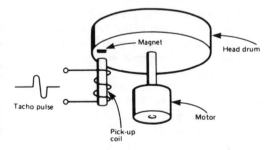

Figure 21.1 *Basic tacho arrangement with rotation magnet and pick-up coil*

299

Hall-effect devices are used. However it is made, each time this *tacho pulse* appears we know the head drum is at a certain angular position.

The tacho pulse, then, may be regarded as a phase reference for the head drum. If the tacho pulse is applied to a phase detector for comparison with the timing of the incoming field sync pulse, an output voltage will be produced (see page 265) proportional to the timing error between these two input signals. If this error voltage is applied to the head drum drive motor, the control loop will ensure a fixed relationship between the field sync pulse and the angular position of the head drum – yet another example of a PLL, but this time with a mechanical system within the loop. This simple servo system is illustrated in Figure 21.2. By delaying either the field sync pulse or the tacho pulse on its way to the phase detector, we can set up any desired phase relationship between incoming field sync and video head position, governed by the length of the delay introduced. It is arranged that the phase relationship is such that the video tracks are laid down with all the field syncs at the bottom of the tape, as in Figure 17.16. The fact that one 20 ms TV field is recorded by each head during one half-turn of the head drum precisely dictates the rotational speed of the drum – it comes out at 40 ms per revolution, or 1500 rpm.

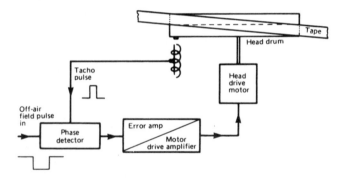

Figure 21.2 *Head drum servo loop during record. The head drum phase (angular timing) is locked to off-air field syncs*

What is required of the capstan servo during record? As we have seen, it takes no part in determining the timing of each one-field track, which is wholly the province of the head drum. The capstan speed governs the spacing of the video tracks, however, and as we know, they are laid down so that they just abut each other; this calls for a steady and even flow of tape past the head drum. The audio signal is recorded longitudinally on the tape in the normal way, so good sound and jitter-free pictures depend on a steady capstan speed. The actual capstan speed depends on the format, being 2.34 cm/second in VHS, for example. During record, then, the capstan needs only to be held at a constant speed, and in some early VTR designs no capstan servo was used. All current VTRs employ a capstan servo, however, which on record is slaved to a high-stability frequency reference known as a *clock*. This may be a stable crystal oscillator or the field sync from the signal being recorded.

Figure 21.3 *Configuration of tape track pattern as a result of the action of the servo loop in Figure 21.2*

The control track

When the video tape being recorded is played back, we will need a reference signal to define the physical position of each recorded video track on the tape. The situation is akin to that of cine film, where sprocket holes are provided to ensure that each frame is exactly in position in the film gate before the projector's shutter is opened. Our video tape 'sprocket-holes' take the form of a control track recorded along the edge of the video tape by a separate stationary head just like the sound head. The control track consists of a 25 Hz squarewave pulse train. One pulse, then, is recorded for each two fields and the track pattern is shown in Figure 21.3 which also shows the position of the sound track. We shall see later how the control track is used during playback.

Servo operation during playback

Let's assume we load into the VTR a good pre-recorded tape, and initiate the replay mode. To get a stable and noise-free picture we need the same close control over the capstan and head drum drive that takes place during recording, but the control references are somewhat different. We'll take the operation of the head drum servo first. The head-changeover point, which needs to correspond with that used during recording, is controlled during replay by the head drum tacho pulse, which triggers the head-switching bistable via a variable time delay. The delay is adjusted so that head switching takes place at the correct instant, just into each tape track.

The speed of the head drum during replay determines the line and field frequencies of the off-tape video signal. Thus the drum needs to spin at precisely 1500 rpm as during record, and this is achieved by slaving it to a stable frequency reference – again a crystal reference is used. Phase-locking takes place in the same phase detector as was used during record; the positional feedback input to this is again the tacho pulse from the drum pickup coil, but the reference input to the phase detector now becomes the local high-stability frequency source – a crystal oscillator as shown in Figure 21.4.

During replay we have to ensure that each video head scans down the centre of its intended track, and this is achieved by adjusting the lateral position of the tape around the head drum to line up the head scans with the recorded tape tracks, and

Figure 21.4 *Head servo on replay, locked to stable local reference*

hold them constantly in alignment with each other. This calls for another servo operation, and a moment's thought will show that this 'lateral lining-up' procedure can be carried out by adjusting tape position (via the capstan servo) or head drum position (via the drum servo). In either case we need a reference signal related to the position of the video tracks on the tape, and this is where the control track, laid down during record, comes into use. We'll describe the capstan-tracking system, as used in most of the early machines.

In addition to its duty of pulling the tape through the machine at a steady speed equal to that used on record, the capstan is now required to position the tape relative to the head drum to achieve correct tracking. Although the video tape tracks are very narrow, they are at such a shallow angle (say 5°) to the tape path that a relatively large lateral movement is required to 'move' the tape across from one track to the next. Let's see how this drum/tape-position phasing is effected.

During playback the control track head reads the 50 Hz pulse train off the tape to provide a 'track position' reference signal for the capstan servo. This is compared in a phase detector with the same stable local signal (50 Hz mains or the frequency-divided output of a crystal oscillator) to which the head drum is slaved, and any error is fed to the capstan motor to pull it into lock. Thus a fixed relationship is set up between drum position and tape track placement, and the phasing, or timing, of this relationship can be adjusted by including a variable delay in the path of the control track signal to the phase detector. This variable delay is set by the tracking control, and it is adjusted on playback for minimum noise in the picture, signifying optimum tracking. A diagram of this system appears in Figure 21.5, while the effect of adjusting the tracking control is conveyed in Figure 21.6. Some VTRs carry out the replay tracking function via the head drum servo.

Automatic tracking systems

So long as the physical position of the sync head is correct, with reference to the entry-point of the tape onto the drum in the recording machine, its tapes will replay in any similar machine, provided that the latter is also in good mechanical alignment. The tracking control only operates during replay, and is only required

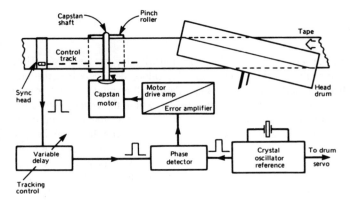

Figure 21.5 *The capstan servo during replay, locked to control track pulses*

Figure 21.6 *The effect of the tracking control on replay. It 'phases' the lateral tape position to line up video tracks with the path of the vision head*

to take up tolerances due to mechanical wear, tape tension variations etc. It is a user control, and maladjustment often impairs picture performance.

Various forms of automatic tracking system have been developed to overcome this problem: we have examined the ATF arrangements of the Video-8 format. With this no manual tracking control is necessary or provided since 'tolerance' errors, both long term and short term, are automatically taken up. Any errors due to tape-path anomalies between record and playback machines are similarly removed. In VTRs using longitudinal control tracks a form of automatic tracking control is possible. The off-tape FM video carrier's *amplitude* is independent of signal content, depending only on the efficiency of tape-head transfer. This means that tracking accuracy can be measured by monitoring the replay FM 'envelope' amplitude. Auto-tracking systems operate by varying the phase delay (bottom left-hand corner of Figure 21.5) to maximise replay head output. This can be done on a continuous or a 'stab and store' basis.

Servo basics

We have seen what servo systems do in a VTR and why they are necessary. Time now to find out how they work! All servos are closed-loop devices, where the 'end result' is fed back to modify the 'processing'. Perhaps the simplest example of a closed-loop system is an electric immersion heater, where the water temperature is monitored by a thermostat which switches the heating element on and off to maintain a reasonably constant temperature regardless (within the capabilities of the heating element) of the demand being made on the hot water supply. A more scientific example of a closed-loop system is found in a TV receiver where incoming line-synchronising pulses are used to hold in step the horizontal scanning lines of the picture. A sample of output frequency and phase is fed back to the flywheel sync circuit and its timing compared with that of the incoming sync pulses. Any discrepancy gives rise to an error signal, and this, after filtering, controls the frequency of a VCO. The local frequency and phase are pulled into lock with the controlling, or reference pulses, and the oscillator is then said to be slaved to the reference. In discussing colour-under systems and colour crosstalk compensation processes we met the closed loop (PLL) system several times, and the phase discriminator and VCO at the heart of it are discussed on pages 265-6.

Figure 21.7 shows the basic servo loop. In this case the motor is a permanent-magnet DC type, so that its speed is proportional (for a given mechanical load) to armature current. On its lower shaft is a disc with a magnet attached so that one pulse is induced in the tachogenerator coil for each revolution of the shaft. This pulse, at 25 Hz, is timed and shaped in the pulse amplifier and then fed to the phase detector. A second input to the phase detector forms the reference, and in our example it is derived from the field sync pulse of a video signal via a sync separator, divide-by-two counter and monostable delay, whose function will be explained shortly. Assume the DC coupled motor drive amplifier is arranged to have a standing 5 V DC at its output with zero input voltage, and that the motor is

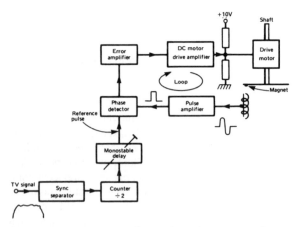

Figure 21.7 *Basic servo loop. The reference signal is incoming field sync pulses*

designed to rotate at about 1500 rpm with such an input voltage. This rotational speed will give rise to tacho pulses at 25 Hz, and if these are coinincident in time with the reference pulses, the phase detector will provide zero output and the motor will continue to rotate at 1500 rpm. Now, if the mechanical load on the motor is increased, its speed will tend to fall, and the phase detector will start to see a time-lapse between the arrival of a reference pulse and that of the tacho pulse, giving rise to an error voltage into the DC amplifier. This error voltage then drives more current through the motor and so increases its speed to restore the status quo. When reference and tacho pulses again become coincident in time, steady-state conditions are restored in the servo loop. Should a speeding-up of the motor occur, the circuit operates in the opposite way to correct this.

Sample and hold circuit

Because of the relatively low pulse repetition rate involved in VTR servos, and the degree of precision required of them, a special type of coninincidence detector is required, known as the sample-and-hold phase detector. A block diagram to illustrate the principle is shown in Figure 21.8. The feedback pulse from the tacho-generator is made to initiate a linear voltage ramp, or sawtooth, symmetrical about

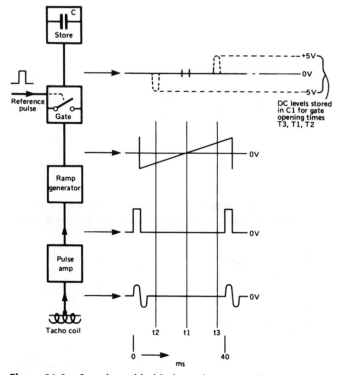

Figure 21.8 *Sample-and-hold phase detector with timing waveforms*

305

zero potential as shown in the waveform. This ramp is applied to a gate (in the form of an electronic switch) which is normally held closed. When the reference pulse comes along it momentarily opens the gate and permits a *sample* of the ramp to charge the sampling capacitor C. Let's assume the reference pulse appears at the instant when the ramp is passing through zero volts, i.e. at time t1. When the gate is briefly opened at this time C will 'see' zero volts and acquire no charge. If the ramp and reference drift apart in phase so that the gate is opened at time t2, C will briefly see a negative potential, to which it will charge. Similarly, if the reference pulse comes late (t3), C will see a positive potential when the gate opens. Provided the discharge path or load on C is sufficiently high it will retain its charge between gate pulses, and that charge will indicate by its polarity the *direction* of the timing error, and by its magnitude the *amount* of the timing error. By amplifying and smoothing the voltage charge of C we shall arrive at an error voltage which can be used to control the speed and phasing of a drive motor.

If the circuit is going to have a symmetrical operating range, the ramp needs to be linear. The simplest way to generate a ramp waveform is to allow a capacitor to charge towards a fixed potential via a resistor as in Figure 21.9a. The problem with this simple circuit is that as the capacitor acquires charge, the voltage across it increases, leaving less for the resistor. The current in a resistor is proportional to applied voltage, so that the capacitor-charging current decays with time, giving rise to the non-linear ramp shown in Figure 21.9b. To ensure a straight ramp it is necessary to charge the capacitor at a constant current, and this can be achieved in two ways.

The simplest is to charge the capacitor from a high voltage source via a high-value resistor. Provided the charging process is stopped (by closing a switch across the capacitor) before the capacitor has charged to a small fraction of the total voltage, the ramp appearing across it will be linear enough for practical purposes. The second method of constant current charging is by means of a feedback circuit, of which the *Miller Integrator* and the *Bootstrap* circuit are the foremost examples, Let us examine the latter type.

Bootstrap configuration

A basic bootstrap circuit appeas in Figure 21.10. The ramp capacitor C1 charges towards the positive voltage line via R1 and R2. In doing so it raises the base

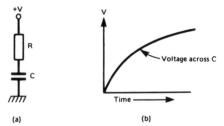

(a) (b)

Figure 21.9 *Ramp generator: (a) shows a simple CR charging circuit, and (b) the resulting exponential charging curve*

Figure 21.10 *The bootstrap ramp generator. This circuit can generate a linear ramp*

voltage of emitter-follower Tr1 and a replica of the charging ramp appears at Tr1 emitter, now at low impedance due to the current amplification of the transistor. The 'followed ramp' is transferred to the top of R2 by the large capacitor C2, so that the charging resistor R2 has a similar ramp waveform at each end. As a result the voltage across it does not change as C1 charges, and the current through it is constant. A constant charging current makes for a linear ramp, and this is taken off at low impedance at Tr1 emitter. To start the ramp, the switch across C1 is opened; when the switch is closed the ramp will be terminated. In practice this switch is a transistor, turned on and off by the tacho pulse.

There is no reason why the tacho pulse should initiate the ramp and the reference pulse open the gate in a sample-and-hold detector, and the roles can be reversed. The important thing is that the sampling capacitor will store a voltage proportional to the phase error between the two. In practical circuits both configurations are used.

Monostable delay

As we have described, the phase detector in a servo system will operate to achieve phasing of a rotating shaft to a reference pulse. If we need to adjust the phase relationship between the two, we must introduce a variable time delay in the path of one of the pulse inputs to the phase detector.

Because the pulse in either case is a timing reference rather than a specific waveshape, it is not necessary to use an analogue delay line; a monostable transistor pair can be used to produce an output pulse at any desired time after triggering pulse, the time delay being governed by an RC time constant, as in the circuit of Figure 21.11. Tr1 is normally off due to its having no base current feed. The arrival of a positive pulse at its base will turn it on, robbing Tr2 of base current. Tr2 now switches off and its high collector voltage pulls up both plates of C1, sustaining base current in Tr1. C1 will begin to charge via R1, and when the

Figure 21.11 *Monostable circuit for pulse delay. The delay depends on time-constant C1, R1*

voltage across R1 has decayed to the point when base current can no longer be sustained in Tr1, the circuit suddenly reverts to its stable state with Tr1 off and Tr2 on. The negative spike generated at this moment on Tr2 collector forms the delayed output pulse and if necessary can be inverted to form a positive pulse. By making part of R1 variable the monostable time delay can be adjusted, and the servo phasing with it.

Motor speed control

The drive motors used for head drum and capstan may be of several different types. Early VTR models favoured a miniature precision DC motor with permanent magnetic field. Its speed is proportional to armature current and for this reason it lends itself well to servo control. Another early approach was to use a large mains-powered synchronous motor with a design speed slightly faster than that required. Here the servo output stage controls DC current in a braking coil whose magnetic field embraces a copper or aluminium disc on the motor shaft. Eddy currents, proportional to the current in the braking coil, load the motor and reduce its speed. The system was heavy in terms of power consumption and weight, and is now obsolete.

A third alternative is the HF synchronous motor. This AC type rotates in synchronism with the frequency of its drive voltage; and for servo control would be fed from the amplified output of a VCO under the influence of the error voltage from the servo's phase detector.

Far the most popular now are direct-drive motors. Here the rotor consists of a disc magnet into which are 'printed' a number of N and S poles. The stator is formed by a number of coils (usually six) arranged in a radial pattern, and fed by an IC multiphase driver circuit. The current-pulse timings are governed by Hall-type magnetic sensors, and speed/torque by stator coil current, in turn controlled by the voltage applied to the drive IC, often built onto the motor itself. Figure 21.12 shows the construction of the direct-drive motor.

Figure 21.12 *Direct-drive motor for capstan drive in a VTR. On the left is the stator with three drive-coil pairs, PG pick-up head and drive IC; on the right the printed-magnet rotor whose shaft forms the capstan proper*

Motor coupling

First-generation home VTRs used belt drive to the 'operating' shafts of the capstan and head drum. Additional jockey wheels and slipping clutch arrangements were provided to drive the tape reel turntables from one of the servo-controlled motors, not only during forward mode but also during fast forward and rewind operation.

Current machines tend to use direct-drive DC motors in which the capstan and head-drum spindles are actually the motor shafts. This greatly simplifies the deck layout and also reduces maintenance by eliminating wear-prone belts and friction wheels. Closer control and greater versatility are also possible with this approach, which can involve up to five purpose-designed motors to drive separately the head drum, capstan, tape reels and threading mechanism. In front-loading machines yet another motor may be employed for cassette transportation.

Practical servo circuits

Current VCRs of all formats use ICs in their servo departments, and the circuit diagrams for these are little different to the block diagrams we have studied so far. To illustrate the working of a ramp generator and sample/hold circuit we must turn to earlier designs, one of which appears in Figure 21.13. Here we have a reference signal, in the form of a 25 Hz rectangular wave, entering on the left. During

Figure 21.13 *Ramp generator and sample-and-hold detector for VTR servo*

each positive half cycle D1 is conductive and effectively short-circuits the charging capacitor C1. For the 30 ms or so duration of the negative portion, however, D1 is off and C1 is allowed to charge via R1. Tr1 and C2 provide bootstrap action and a linear ramp is thus generated at point A, whose slope is adjustable with VR1. The 'ramp down' waveform appears at point C, where it is grounded in T2. This transistor acts like a switch, normally held on by base bias via R6.

The tacho generator coil at the bottom of the diagram generates one pulse for each revolution of the head drum or capstan, as the case may be. This pulse is shaped and amplified to appear at short-duration negative spikes at C3. Each time one appears, Tr2 is momentarily switched off, and its shunting effect on the ramp waveform removed; a sample of ramp voltage *at the instant of the gating pulse from the tachogenerator* is allowed to pass onto the bases of Tr3 and Tr4. If this sample is higher than Tr3 emitter voltage, the transistor will conduct and charge the storage capacitor C4 to the sampling level. If, however, the requirement (due to direction of phase error) is such that charge needs to be 'pulled out' of C4, Tr3 and D4 will remain reverse-biased and non-conductive, but Tr4 will come on to discharge C4 to the new sampling level.

For C4 to maintain its charge between sampling pulses it needs to be looking into a high impedance, and this is provided by the compound emitter follower Tr5/Tr6, the output of which (point D) forms the error signal.

The junction of the two 5K6 resistors R4/R5, point B, is the measuring point for the ramp and sample circuit. An oscilloscope connected here will show the sample pulse sitting on the ramp, and if the circuit is correctly adjusted the pulse will appear half-way down the ramp as in Figure 21.14a. This will offer a symmetrical correction range. If the servo is out of lock, due perhaps to a fault condition, the pulse will appear to run up or down the ramp at a speed propor-

(a)   (b)   (c)

Figure 21.14 *The waveform at point B of Figure 21.13: (a) correct operation; (b) servo unlocked, pulse running 'backwards' up ramp; (c) pulse jittering about correct point due to mechanical problems*

tional to the frequency/rpm error, as conveyed in Figure 21.14*b*. Mechanical jitter, due to worn bearings or faulty friction drives, causes the pulse to jitter about the correct point on the ramp, as in Figure 21.14*c*.

Motor drive amplifier

The amplification system between the error generator and the drive motor needs to be DC coupled, and have high gain and good stability. Typically a motor requires (under normal loading conditions) between 40 and 120 mA at 6 V, but the drive amplifier must be capable of supplying a much greater current than this to ensure quick acceleration from rest, and particularly to ensure rapid response in the motor to a change in drive from the servo system during normal phase and speed correction within the loop system. Implicit in this is the necessity for high 'loop gain' and we'll return to these characteristics later.

A discrete design of servo and motor-drive amplifier is the best instructional model (see Figure 21.15). The capstan servo is shown, and the rotational speed of the capstan is such that its tacho pulses appear at 3.71 Hz rate, much lower than the 25 Hz rate of the head drum pulses. In most other respects, the drum and capstan servos in this machine are similar. The upper part of the diagram is the phase detector, taking a positional feedback signal from the capstan tachogenerator and comparing it with a crystal reference (2.51 MHz, divided to 3.71 Hz) in the servo chip IC4. The error signal is developed on C39, and appears across C41 on pin 16 of the chip whence it is added, via R61, to a standing voltage of about 6.3 V from the potential divider chain R65/66/67. The potentiometer R66 sets the DC operating conditions for the motor drive amplifier and is adjusted for correct 'free running' motor speed. This has the effect of restricting the servo-control effect to a narrow 'window' centred on 6 V or so, and ensures that the motor quickly runs up to speed from rest aided during the play mode by the charging effect of C50 via D12. Motor control voltage is fed into the non-inverting input of an operational amplifier IC5. The output from pin 7 of this chip passes to a Darlington pair X12 and X2, with the DC motor forming a load for X2. X13 forms a shunt switch operated by the mechanism control circuit. When pins 1 and 2 of the inverter IC3 go low, its pin 3 will go high, turning on X13 and grounding the base of X12. As a result X12, X2 and the motor will turn off.

Figure 21.15 Capstan servo circuit with IC discriminator and discrete-transistor motor drive (Ferguson)

A *discriminator-feedback* system is used in this motor drive amplifier, in which the reverse EMF (generator effect) of the motor is used to control the op-amp IC5. Motor current (proportional to load rather than speed) is monitored by 0.47Ω resistor R81, and the voltage it develops is in effect applied across the inverting and non-inverting inputs of IC5, which then operates to balance motor voltage and control-input voltage. In addition, back EMF from the motor, appearing whenever its mechanical load decreases, is passed into IC5 as negative voltage feedback to its inverting pin 6, reducing motor drive to match the reduced loading on the motor.

Loop transient response and damping

The design of a servo system is critical if its performance is to be good, and VCR servo systems are very demanding in this respect. The performance of a servo is traditionally analysed in terms of a step input function, in which a step, or instantaneous change, in control voltage is applied and the output function (in our case, capstan or head drum phase position) analysed to see how closely it corresponds to the step input.

This is illustrated in Figure 21.16, where the step input is shown along with three possible responses at the servo output. Plainly it is not possible to achieve an instantaneous change in servo output where a mechanical system is involved – the inertia of the moving parts will cause an inevitable time lag before the new phase condition is established, and the designer's aim is to do this in the shortest possible time without spurious mechanical effects. Three factors are involved – damping, gain and time-constant – and each of these are present in mechanical and electrical form. An ideal servo system has each of these factors optimised, and the electrical design of the motor drive amplifier takes into account the mechanical characteristics of the motor and shaft coupling systems, and inertia of the rotating parts. Electrical analogies of mechanical functions can be built into a circuit in such a way that deficiencies are catered for – for instance, positive feedback via a capacitor will reinforce the gain of a motor drive amplifier only when the *rate of change* of drive amplifier output is high; the same effect can be realised by negative feedback via an integrator. This may be used to overcome the inertia effect of a heavy flywheel. Very often two time-constants are used to control the response time of the servo loop: one to quickly achieve correct speed (frequency) and a second to establish exact positioning (phase).

Figure 21.16 *Loop transient responses compared. Waveform (a) is the step input; the other waveforms show possible reactions at the drive shaft*

Let's now look at the servo responses illustrated in Figure 21.6. The step input is shown in curve (a). Curve (b) illustrates the effect of excessive damping, in which the output takes a long time to follow the input signal resulting in sluggish operation; curve (c) shows insufficient damping, causing the servo output to *hunt* or oscillate about the correct point for a period before settling down. Without any damping, continuous oscillation would take place! The ideal situation is shown in curve (d), that for *critical damping*. This is the one which designers normally aim for as offering the best compromise between response speed and hunting effect. A single slight overshoot and one small undershoot is followed by stable operation at the new level. This effect is easily demonstrated in most machines by making a sudden and well-defined adjustment of the tracking control to simulate a step input function to the servo control. The picture will be seen to move sideways in one or two damped cycles of oscillation before settling down.

There is great variation in the design of motor drive amplifiers, depending on the mechanical system and intended conditions of operation of the machine. Much depends on motor design, and the servo system we have described for the early Ferguson machine is one of the simplest in use. Other contemporary designs were more sophisticated, and later developments like the portable machines now available, have more complex feedback arrangements in their servo systems. The development of brushless multipole DC drive motors with purpose-designed IC control systems has greatly improved servo performance in current VTR designs.

Digital servo system

An alternative approach to servo comparator design is embodied in the digital system illustrated in greatly simplified form in Figure 21.17. Instead of the analogue ramp-and-sample arrangement, a counting system is used to derive an error voltage proportional to the time delay between reference and sample pulses. A clock oscillator runs at a fixed frequency to operate the system, and the arrival of a reference pulse *enables* counter 1, which starts to count clock pulses, feeding them into a memory. The memory is 'frozen' each time a sample pulse appears, and at that moment in time the stored count is proportional to elapsed time between the reference and sample pulses. We now need to compare this stored count with a reference, or time marker, and this appears in the form of counted-down clock pulses from counter 2. These are compared with the memory contents in a comparator whose output is converted to analogue form in a D/A converter consisting of a bistable and charging capacitor. The error signal thus produced controls the drive motor in the normal way, and the circuit has the advantage of eliminating the presets and manual adjustments of the conventional sample and hold discriminator.

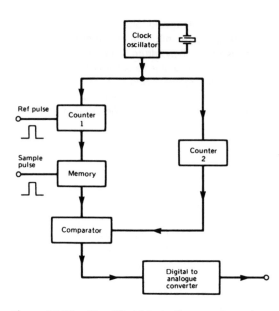

Figure 21.17 *Simplified block diagram of one form of digital servo*

Still-frame requirements

Chapter 17 related how various machines solve the problem of noise-free freeze-frame reproduction by means of four-head drums or extra-wide replay heads to overcome the change in effective tape track angle between moving and stationary tape. When the tape transport is stopped at a random point, track scanning by the vision heads may be such that a mistracking bar is seen on the still frame.

It is necessary to position the tape so that this noise bar is 'shuttled' out of the picture to give satisfactory reproduction. This is carried out by the capstan servo which 'inches' the tape forward until the noise bar moves out of the picture into the field blanking period, when capstan drive ceases. The reference for the inching system is off-tape video, usually a sample of FM luminance from the DOC circuit, though certain early-design machines use the sampled phase error of the replayed chrominance signal to achieve the same result. The presence of the noise bar in the field blanking period would tend to obliterate the field sync pulse during replay, and even it it were available, the half-line 'jump' in its timing between successive fields would cause picture judder on still frame in machines where this system is used. For these reasons, an 'artificial' vertical sync pulse generator is used in still-frame mode, and its pulse timing is governed by a preset, adjusted for minimum vertical judder of frozen pictures.

An extension of the inching idea outlined above makes possible slow-motion replay in suitably equipped machines. Let's assume we need to replay at one-third normal speed. It will be necessary to position the tape accurately on the head drum to achieve good still-frame reproduction as described, and hold it in position

315

for three sweeps of the video heads, i.e. three revolutions of the drum. After this the tape is quickly moved forward by the capstan to line up the next track accurately for the next three revolutions of the head-drum, and so on. The same process can be carried out with the capstan going backwards to give a slow-motion sequence in reverse. The precision required during these processes is of a high order and the circuits to carry out the operation use a mixture of analogue and digital techniques, often with specially designed ICs. A description is beyond the scope of this book!

Tension servo

During record or playback, it is important that the tension on the tape is held at a constant level, regardless of the amount of tape on each spool and the friction encountered by the tape on its journey through the deck. Insufficient tension in the tape during its wrap round the head drum leads to poor head-tape contact and results in poor signal transfer and noisy pictures. Too great a tension causes head and tape wear and tape stretching, and can result in poor compatibility with other machines due to 'stretch-distortion' of the magnetic pattern of the tape. When replaying a 'foreign' tape on a machine with excessive back tension, the registration between luminance and chrominance signals is lost, and the colour will be 'printed' to the left of the luminance image in the reproduced picture.

To prevent these problems, VTRs incorporate a tension servo to maintain a constant 'stretch factor' in the tape. In simple machines this takes the form of a passive mechanical negative feedback system, in which a tension arm bears against the tape as it leaves the feed spool in the cassette. The arm is linked via a lever to a brake band wrapped around the supply spool turntable, so that low tension causes the brake band to bind against the supply turntable and vice-versa, maintaining reasonably constant tension in the tape. The arrangement is shown in Figure 21.18.

Figure 21.18 *Simple brake-band tension regulator*

In machines where each tape spool is driven from its own motor, a more precise control of tension is possible. Reverse current in the supply-spool motor sets the back-tension in the tape. Typically the tape emerging from the cassette passes over a spring-loaded feeler which carries a shutter; this is in the light path of an optocoupler (LED-phototransistor combination) so that conduction in the phototransistor is proportional to tape tension. The phototransistor current is amplified and used to control the current in the *supply* spool motor, which is arranged to run 'backwards' at this time. Because of the steady pull on the tape from the capstan, the supply motor will be pulled in the opposite direction to which it is trying to run, and the motor current will then determine tape tension under the control of the optocoupler sensor. At the beginning of a cassette, when all the tape is on the supply reel, the reverse current in the supply spool motor is of the order of 250 mA. As the tape is wound on to the take-up reel this current drops to maintain correct tension, as illustrated in Figure 21.19, which also shows the constant forward current in the take-up spool motor, representing the latter's steady pull on the tape away from the capstan and pinch roller.

Figure 21.19 *Electronic tension servo: motor-control currents vary with the amount of tape on each spool*

22 *System control for VTRs*

Several times in this book, we have compared aspects of audio and video tape recording and the machines for doing so, and in each case we have seen how much more sophisticated is the video system. Simple audio tape recorders have only a single motor; no electronic circuit or motor design will make it rotate in opposite directions at the same time! All that is necessary, then, is a simple mechanical interlock to prevent more than one transport button being depressed at a time. In high-quality audio machines with fast spooling and more than one motor, a simple form of system control is provided. Once again, then, in the area of system control, the VTR has a multiplicity of requirements which do not exist in its audio counterpart. These arise from: the threading in and out of the tape from the cassette; the high rotational speed of the head drum; the fragility of the vision heads; the use of several drive motors whose efforts could conflict; the high speed and torque of the fast-wind and rewind processes; the need to control the machine by light-touch sensors or a remote keypad; and several minor factors.

Need for system control

The system control (we'll call it syscon for short) section of a VTR may be regarded as a 'policing' system for the mechanical functions of the machine. It has two basic interrelated functions, preventing user 'abuse' and programming the machine's mechanical operations, while protecting VTR and tape in the event of machine malfunction or tape faults. The syscon receives inputs from the user's mode keys (command functions) and also from monitoring points around the tape deck (protection functions). These are correlated in the syscon and so long as no conflict exists between command and deck or tape status, the mode required is permitted. If at any time during any mode, a relevant protection input is received (e.g. drum motor stops during play or tape-end sensor is activated during rewind), the syscon will invoke a new mode – 'stop'! The syscon works in digital mode, and recognises on each input line only two states, on (1) and off (0). It presents its outputs to solenoids, motors etc. in the same way. As in previous chapters we will use 'discrete' circuits for illustration.

Syscon inputs

Let's examine the inputs to the syscon in turn. The command inputs are the user control keys – play, record, rewind, fast foward, fast rewind, pause, stop and, where provided, forward and reverse picture search. The tape counter also provides a command function (stop) if it reaches 0000 or a pre-programmed number ('go-to' facility) with its memory key depressed. The protection inputs are many and varied, and we'll look at each of them, and the effect they have.

End sensors. The tape-end sensors are activated by markers 150 mm or so from each end of the tape: an LED shines through a transparent leader tape to activate a photo-cell.

The tape-end signal thus derived invokes stop mode and then prevents any further tape transport in the 'unsafe' direction. Many machines also incorporate a protection circuit which monitors the current in the tape LED; if an open-circuit here went undetected no tape-end signal could be given, and damage might result.

Rotation detector. We have seen that the head drum generates tacho pulses as it rotates, and their cessation would mean that the head drum is stalled. This is monitored by the syscon, along with tacho pulses from the take-up spool carrier. If either fail, stop is entered after a short time. During freeze-frame or picture pause modes both tape spools will stop, so the spool rotation detector output is over-ridden in these modes. To prevent tape and head wear, a timer is started which enters stop after a few minutes' scanning time over a single tape track.

Cassette-in. All modes are inhibited when there is no cassette in the machine. A lever and microswitch detect the presence of a cassette, being deflected by its plastic envelope as the cassette carriage is driven home into the machine.

Recording tab. As with audio cassettes, video types have a removable safety tab to prevent accidental erasure or recording-over the programme it contains; where the tab has been removed, a lever and microswitch detect the fact, and signal the syscon to prevent record mode being entered.

Threading completion. Sometimes known as the after-load (AL) switch, this closes when the tape is fully threaded around the head drum. Until threading is complete, the syscon will inhibit capstan drive to prevent forward motion of the tape. *Mode-switches* give more comprehensive information.

Dew sensor. If the ambient temperature in which the VTR operates changes suddenly, usually due to bringing it indoors from location work, there is a risk of vapour condensation, particularly on the head drum. If the machine were allowed to operate in this condition the tape would stick to the head drum with disastrous results! To detect condensation, then, many machines are fitted with a dew sensor on the stationary part of the head-drum assembly; it consists of a pair of metallic fingers or a high-resistance 'element'. The presence of moisture increases the resistance of the sensor, and this is monitored by the syscon which acts to enter stop and illuminate a warning lamp. Some machines incorporate a low-wattage head heating element to drive off condensation. These features are important in portable equipment, and are present in many 'stationary' home VTRs.

Power interruption. When the machine is running in any normal mode, the syscon is in a 'latched' state, so that interruption of mains power will unlatch the circuit. The result, in most machines, is that stop will be entered on restoration of power.

Tape duration measurement. Sophisticated machines have a 'time-elapsed indicator' as a front-panel readout, and to operate this it is necessary to know what the cassette type is (2 hour, 3 hour etc.) and how much tape is left on each spool. Some cassettes have a code embossed into the cassette body to indicate running time, and this operates sensors in the VTR; some sophisticated machines of various formats compute spool speeds relative to each other against the fixed tape speed to arrive at a figure for cassette type in hours'-worth of tape. Once the tape running time is known, it remains to determine the amount of tape on each spool. This may be carried out via the tension servo system in some designs, or by an extension of the spool-speed-versus-tape-speed computation described above.

Syscon outputs

Motor stop. The capstan and drum motor can be enabled or disabled by on-off signals from the syscon. These usually operate transistor switches in the motor drive amplifiers – an example is X13 in Figure 21.15.

Pinch roller solenoid. This pulls the pinch roller into engagement with the capstan shaft to enable the latter to pull the tape through the machine. It will disengage during pause mode. In many VTR models fast tape transport takes place with the tape laced up, and the pinch roller will be disengaged in fast-forward and rewind modes.

Reel brakes. In some VTR designs, tape reel brakes are operated mechanically from the loading motor cam. Other designs use magnetically applied brakes under the control of the syscon. During the threading phase the spools need to be unbraked to enable a loop of tape to be drawn out of the cassette. During unthreading, spool drive needs to be applied to take up the tape loop, pulling in the slack as it becomes available.

Indication lights. The syscon drives panel-mounted LED, fluorescent, or on-screen displays to indicate machine and tape status. The number and sophistication of these depends very much on make, model (and price!) of the VTR.

Programming muting. To prevent unlocked pictures and mistracking effects from being seen as the machine runs up to speed (*lock-up time*) the VTR's video – and sometimes audio – sections are muted, either for a fixed time-period after play mode is entered, or until the servo systems are locked up. In the latter case, the mute will operate in the absence of off-tape tracking pulses (e.g. blank or faulty tape), and in the event of servo malfunction. During 'trick-speed' operation, such as pause, shuttle search and slow or reverse motion, the servos will not be operating normally, and the syscon must override the auto-mute circuit.

Syscon circuit

We have seen that all inputs and outputs from the syscon are in binary form, or logic levels 0 or 1, and that the syscon function is one of decision making. This means that a syscon can be made entirely of logic gates, inverters and counters, though interfacing circuits are required to enable their outputs to drive solenoids, motors, brakes and so on.

First-generation VTRs used transistors and diodes in their syscon circuits, and the requirements lent themselves to the use of simple digital IC arrays such as 74-series TTL chips. The next logical (!) step was to integrate all logic functions into a single chip containing the necessary gates and counters for a particular machine's requirements, and this idea was adopted in many early-design machines in all formats. A syscon of this sort is illustrated in Figure 22.1, together with its inter-facing circuits. The internal workings of the chip are fairly obvious from the diagram, but it is instructive to look at its peripheral circuits and interfaces, as they give a useful insight into the overall operation of the machine and an intro-duction to more sophisticated syscon designs in later machines. We'll examine circuit operation from the point of view of each IC pin.

Pin 1: 'Operate' in. Goes high in all modes except stop, and primes the chip for further commands.

Pin 2: Rewind in. High when rewind is selected.

Pin 3: Play in. High when play key selected.

Pin 4: Pause in 2. Fed from X11; high during pause mode to initiate 5-minute pause counter within the chip. At the end of the count, stop mode will be entered.

Pin 5: Pause output. High during pause; feeds pin 4 and pause rail out via X11.

Pin 6: Motor control. High during drive modes; this is turned to a low in inverters IC3 to switch off motor control transistors X12 and X13, thus enabling drive motors.

Pin 7: Flip-flop in. Drum rotation detector input, in the form of head drum tacho flip-flop pulses. Their absence will trigger stop mode after 5 seconds, timed by the stop counter within the chip.

Pin 8: Record in. Not used in this machine, and held permanently at 12 V (high).

Pin 9: Pause in 1. High when pause key down, and in stop mode. Releases pause solenoid to stop tape transport.

Pin 10: Pause solenoid output. Goes high during pause mode; this turns on X10, and also X8 and X9 via C11. X9 draws a heavy current through the upper winding of the pause solenoid from the 18 V supply and the armature pulls in. Within a second or two C11 has charged and X8/9 turn off, but the armature is held in by the much smaller sustained current from the 12 V line through the lower ('hold') solenoid winding and X10.

Pin 11: Remote pause in. Takes command from a camera or wired remote pause control via rear-mounted jack. Normally grounded, high to pause.

Pin 12: Tape-guard input. Not used in this machine, and held permanently at 12 V (high).

Pin 13: After-load in. When the tape is fully threaded-up, S6 closes, and after a short pause during which C10 charges through R17, X7 switches on, putting a high on chip pin 13, whereupon the pause solenoid engages and tape transport begins. During threading, the motors and servos are operating, so that they are locked up by the time tape transport starts. If no AL signal is received within five seconds of requesting play, the syscon will enter stop.

This machine has facilities for serial recording, which means that it can be timer-programmed to record a series of TV programmes on consecutive days.

Figure 22.1 The syscon circuit of a basic VTR. The purpose-designed IC is surrounded by interfacing circuits for sensors and 'operators' such as solenoids and motor switches (JVC)

During this mode, the tape may be already threaded-up when unattended recording begins, and unless precautions are taken, at the appointed time the motors will start from rest, run up to speed and 'lock in' while recording is taking place. This would result in a very ragged start to the recording, a situation that is prevented by the circuit connected to the base of X7. Under the circumstances described, the AL switch will be closed when recording begins, initiated by the appearance of +12 V at S6. C10 cannot now charge quickly because its charging current is diverted by D5 into the 'empty' C12, so that the charge time for C10 is now tied to the much longer time constant C12/R25. X7's switch-on is thus delayed to enable servo locking to be completed before the pause solenoid is energised to initiate tape transport. Should power be interrupted during operation, D6 rapidly discharges C12 into the decaying +12 V line, ensuring the long time-constant when power is restored. D7 maintains full charge on C12 during modes other than timed serial recording; this ensures that D5 is reverse-biased isolating the long time-constant circuit from X7 base.

Pin 14: IC ground for operating potential.

Pin 15: Stop solenoid out. Normally high, this goes low during stop mode to turn off X4. When X4 goes off, current flows via D1, R9, D2 and R13 to turn on X5 and X6 and energise the stop solenoid. Because the solenoid is automatically de-energised (by the cessation of 'ope in' potential on chip pin 1) when stop is achieved, the stop solenoid has only a single power-winding working from the +18 V supply. This contrasts with the separate 'pull-in' and 'hold' windings of the pause solenoid, which is required to be energised for long periods.

If, because of operator abuse, a function key is held down during 'protection' conditions, the syscon will hold off motor drive (via pin 6) and pulse the stop solenoid at two-second intervals until the key is released; this is the function of the 2 second timer within the chip.

A 'conditional' stop function is provided by the circuit centred on X3, and it governs the stop solenoid operation druing mains power failure. During play or record mode, S2 ('play' key) is closed and X3 collector potential low due to its base 'turn-on' bias via R11. D3 is thus reverse-biased and has no effect on circuit operation. If the mains supply fails, the +12 V line collapses, X3 base bias disappears and it turns off. D3 now starts to conduct and the charge on C9 flows through R16 to maintain base current in X4, holding off X5, X6 and the stop solenoid until all power lines have decayed to zero. Thus the VTR remains 'primed', and normal operation is restored when the mains returns. If power failure should occur during fast-wind or rewind, however, S2 will be open and C9 discharged. X4 will thus be allowed to turn off and the stored energy in C7 (isolated from the decaying +12 V line by D1, now reverse-biased) will provide base current for X5 so that the stop solenoid will be energised. Solenoid energy comes from the large capacitor C101 (isolated from the decaying +18 V line by D101). Enough energy is stored in C101 to pulse the stop solenoid once at the instant of mains failure.

Pin 16: Reset in. At switch-on, all functions in the syscon chip must be reset. This is done by a positive pulse from the differentiator C6/R5 operating from +12 V.

Pin 17: End sensor. A high on this pin will initiate stop mode, and while it remains high the syscon will only permit rewind mode. The cassette lamp shining through clear 'end leader' tape will illuminate end-sensor photo-transistors X101, causing them to conduct via R3 and R4.

Pin 18: Start sensor. As for pin 17, but the high signalled by start-sensor photo-transistors X102 will cause the syscon to inhibit rewind mode.

X1 also operates on chip pins 17 and 18. Normally on, it holds D3 and D4 off by grounding their anodes. If the cassette lamp should fail, X3 will be robbed of base current and will turn off, raising anode potentials on D3 and D4 to put a high on chip pins 17 *and* 18, invoking stop mode and preventing motor drive. The same action occurs in the absence of a cassette, when S12 is closed, grounding X1 base.

Pin 19: Unload in. Operated by unload microswitch S7, which maintains a high on chip pin 19 until tape unthreading is complete, whereupon the syscon switches off drive motors via its pin 6. A second unload microswitch (not shown in Figure 22.1) operates to hold in the 'timer-on' relay until unthreading is complete so that the machine goes to stop at the end of a single timed (unattended) recording.

Pin 20: Take-up sensor in. To detect rotation of the take-up spool a *Hall IC* is used. The device is mounted close to a rotating magnet driven from the take-up spool and responds to the nearby alternating magnetic field by an internal 'deflection' process explained in Figure 22.2. The AC output from the Hall IC is fed via R4 and C4 to chip pin 20 where it is monitored; its cessation indicates that the take-up spool has stopped, and after five seconds delay the syscon will enter stop.

Pin 21: Memory in. Memory in this case means 'tape counter at 0000'. A switch S9 within the counter will close at this point, and provided the operator's memory switch S10 is 'on', a high is presented to chip pin 21; the syscon will enter stop.

Pins 22, 23 and 24: Clock timing. Within the syscon chip is a 20 Hz oscillator for use with the 2 second (stop-cycling), 5 second (stop protection delay) and 5 minutes (pause timer) count-down dividers. The 50 ms period of this clock oscillator is set by R3 and C3 connected to chip pins 22, 23 and 24.

Pin 25: Pause logic in. Unused in this machine, and grounded.

Figure 22.2 *Internal operation of the Hall-effect IC. The flow of electrons through the Hall element on the left is, in the absence of a magnetic field, down the centre. An alternating magnetic field will deflect the electron stream to right and left, rather like an electron beam in a TV picture tube. The alternating charges picked up by the side electrodes are amplified and passed out on pin 3 of the chip*

Pin 26: Record lamp out. Not used in this machine.

Pin 27: Pause select. The pause timer delay is governed by the potential applied to this pin. Current video tapes can withstand scanning of a single track for several minutes, and the potential divider R1/R2 sets 6 V on chip pin 27 to give approximately this delay before invoking stop. The range 0-12 V at pin 27 varies the delay time.

Pin 28: VDD in. This is the operating potential for the logic circuits within the chip itself, +12 V.

Simple syscon summary

The system described and illustrated above is effective, relatively simple and virtually foolproof. Inherent in the design, however, are several limitations: remote control is limited to a single *wired* pause function; machine operation is necessarily via 'clunky' keys which operate many deck functions by direct mechanical means; and unattended recording is limited to a single programme during the three days (or one week, depending on the clock capabilities) following clock setting, or serial recording from one TV channel at a fixed time of each day. The system has the advantage of low cost and economy of drive motors.

Touch button deck control

Perhaps the greatest objection to mechanically operated VTRs was the heavily-loaded piano-key operating controls, and these may be compared with the mechanical tuner buttons of early TV receivers. In TV design the arrival of electronic *(varicap)* tuners paved the way for the provision of electronic station selection via light-touch (microswitch) buttons, or touch sensors in which the mere presence of a finger on a sensitive surface triggered on an electronic latching circuit to select the required channel. The next generation of home VTRs, then, offered light-touch selection of deck functions, but at a cost of much greater complexity in the syscon department, on two counts: the loss of the mechanical interlock facility, and the necessity to provide motors and solenoids to do the mechanical chores previously performed by the operator via levers connected to the piano keys.

The loss of mechanical interlock between selectors means that the operator can enter 'illegitimate' commands such as fast-forward during the rewind process, or play and stop at the same time. This requires the syscon to 'think harder', as it were, assigning priorities and storing user instructions until it has achieved a deck status which permits the newly requested mode. The requirement for the deck to do its own housework, so to speak, multiplied the number of syscon outputs, with spool brakes, spool drive motors, eject solenoid, threading motor and comprehensive indicator LED systems having to be provided and very precisely controlled.

While this could be achieved by an extension of the standard logic techniques of the control chip described earlier, a great deal of circuitry would be required, so an alternative technique is used. This is the microprocessor control system and a syscon using this type of chip is shown in block diagram form in Figure 22.3. Inputs, both command and protection, enter the chip on the left, and outputs leave on the right. A great deal of interfacing is required between the syscon chip and the functional parts of the machine, and IC1801 in Figure 22.3 is supported by eight further chips and a multiplicity of transistors and diodes. The control keyboard has grown up from a series of lever-operated microswitches to a matrix system like that in a pocket calculator – pressing a button links two input pins on the syscon chip and the command required is interpreted within the chip.

Real-time tape counter

The simplest tape counter systems merely count the revolutions of the take-up (TU) spool. This is a very non-linear measuring system because the tape length (and hence running period) between counts depends entirely on the amount of tape which has accumulated on the TU spool during playback and the resulting diameter of the reel. Real-time tape counters provide a reasonably accurate idea of the elapsed time at any point in hours, minutes and seconds. They depend on counting control-track pulses at the CTL head in record and playback modes, reckoning them at the rate of 25 per second and displaying the accumulated count on an LCD, fluorescent or on-monitor-screen readout.

Index search

In Chapter 21 we saw that the control track pulses are written onto the tape primarily for the purpose of servo phase-lock – aligning the paths of the video heads with the centres of the tracks over which they are sweeping. In fact the servo system only uses the leading edge of each pulse induced in the CTL head by the control track. This opens the way to the use of the CTL pulse for a third purpose, that of writing an index mark on the tape. Normally the mark/space ratio of a CTL pulse is 60:40. To make an index mark the ratio is briefly altered to 28:72, which makes no difference to the operation of the servo circuit, but can easily be picked up by an integrator or time-counter to give a 'flag' signal whenever a mark is detected.

At the start of each new recording – and at other points programmed by the user if required – an index mark is automatically put on the tape. During replay the user can invoke index search in either direction, whereupon the VTR will fast-transport the tape as required, reverting to play mode as soon as an index mark is encountered. Developments of this system are *intro-search*, in which the machine will play for a few seconds at each index mark before fast-transporting to the next; and multi-index search, whereby the user can program the VTR to stop and play at the *n*th index mark in either direction. Hence the need for the half-loading mechanics described in Chapter 17. Video-8 formats have similar facilities for both indexing and real-time tape counting, but necessarily rely on track-counts from the video heads.

Figure 22.3 Block diagram of a syscon for 'electronic' tape deck control. This one is built round a microprocessor IC (Sharp)

327

Microprocessor control of full function machines

The basic microprocessor control system in Figure 22.3 caters only for mechanical control of the deck, and may be regarded as an 'electronic' version of the simple syscon previously described. For complete versatility in VTR control and facilities, an extension of the syscon is required to cater for further functions. Complete cordless remote control of all deck modes, including trick-speed effects, was introduced. Multi-event timers, in which different broadcast channels could be programmed for recording at various future times and dates, became available. Front-loading machines appeared with the means of mechanically transporting the cassette into its carrier, then lowering the carrier itself into position on the deck. 'Clever' editing of tapes became possible.

Machines incorporating these features need a very sophisticated syscon akin to a small computer to correlate, organise and carry out all necessary control and protection functions, and our earlier analogy of system-policing now has an additional comprehensive 'housekeeping' function. This is an ideal role for the microprocessor, and these devices are used in domestic VTRs, not only in syscon circuits but in such areas as servos and clock-display electronics.

Basic microprocessor operation

A microprocessor is an LSI (large-scale integration) integrated circuit. It works in quite a different way to such logic chips as the 74 series and the first-generation controller described earlier, in that it functions in similar manner to the CPU (central processing unit) of a computer. In normal computer applications it is supported by one or more separate memory chips, in which data is stored and retrieved by the CPU as required.

The sorts of microprocessors found in VTR machines, however, need very little memory capacity, and it is possible to design a complete simple computer system into a single package, perhaps better termed as in 'integrated microcomputer'. It contains all the elements of a small general-purpose computer on a single chip. Thus (Figure 22.4) we find a CPU (central processing unit), ROM (read-only memory) RAM (random-access memory) *clock*, and I/O (input/output) *ports*. The CPU is the heart of the device, controlling the interchange of information between the various sections of the chip and the I/O ports. The ROM contains control programmes covering all combinations of input signals to the system, and these instruct the CPU when the latter calls for them in turn. The RAM is used as a temporary 'storeroom' for data which needs to be held pending the completion of mechanical or data-processing operations. Input and output ports accept requests and dispense commands to and from the microprocessor. The clock is a steady source of timing pulses which 'steps' data sequentially through and around the microprocessor. Address, data and control bus systems are the internal highways for data interchange between the sections of the chip; data is loaded into these on a sequential basis, preceded by an electronic 'label' to indicate its routing and destination.

Figure 22.4 *Internal block diagram of a microprocessor IC*

Under the control of its internal and pre-programmed 'action memory' the micro can cope with any combination of circumstances, and come up with the most logical answer. Taking our earlier example of the user keying in fast-forward during a rewind operation, the micro-syscon will enter stop, wait for that to happen, then programme the deck for fast-forward operation. In other designs, fast-forward will be entered immediately, with reel motor drive and braking controlled in such a way that the changeover is effected in the shortest possible time consistent with not stretching or 'looping' the tape. If more memory or processing capacity than can be accommodated in one microprocessor is required, then an additional slave chip can be tacked on, in the manner of Figure 22.5.

Figure 22.5 *The versatility of the microprocessor can be increased by the addition of a 'slave' chip*

Multiplexing

In the sorts of machine we've described, it's not hard to imagine that the number of command and feedback inputs, and control outputs from the syscon chip will run to over 100. It is impractical to provide more than 100 or so pins on an IC, so that a time-sharing system, known as multiplexing, is used on input and output ports. This is carried out by means of expander chips. These work on a sequential basis, switching several data lines to and from the micro ports in turn when requested by the control or 'chip-enable' line from the micro's CPU. During the 'waiting' period, the expander chip stores data in its own internal memory.

Key scanning

A further economy in data lines is achieved by the process of sequential scanning of the control keyboard. The keyboard is wired as shown in Figure 22.6, and is fed at the top with 'strobed' scanning pulses of perhaps 1 ms duration in sequence, so that SP1 is energised for 1 ms, then SP2, SP3 and SP4.

The output lines are monitored by the micro for returning pulses, so that if line OPB 'lights up' during SP3, button 10 has been pressed; if line OPD comes on during SP4, button 16 has been selected, and so on. The returning pulse from an OP line is decoded within the micro and stored in its memory as a command to be processed.

This key-scanning system used a total of eight lines to handle 16 keyboard commands. The number of keyboard lines can be further reduced by more

Figure 22.6 *Key-scanning matrix for economy in wiring and IC pins. Operation is explained in the text*

complex scanning systems. Some VTR designs, notably camcorders, have a scanning system embracing not only control keys but such things as deck sensors.

Remote control

For cordless remote control of VTR functions, an infra-red link is used between the VTR and a hand-held commander unit. The latter consists of a keyboard linked to a digital encoder chip which prefaces each command by a *framing code* (which identifies the commander and primes the decoder), and then sends a string of binary pulses called a word, each unique to the mode requested. With only a single infra-red link, the data is necessarily sent in serial form. On receipt in the VTR, a *shift register* converts the serial data to parallel form ('static' data on separate lines) for scrutiny by the keyboard scanning system of the microprocessor or for decoding within a dedicated section of the microprocessor itself.

Many remote control units have a timer setting facility in which the times, dates and channels of the TV programmes required to be recorded can be key-stroked in and displayed on an LCD panel on the hand unit. When they have been checked and verified by the user, the data, which has been stored in a 'long' register within the handset, is transmitted to the VTR in a long string of binary words, prefaced by the usual framing code. In a fraction of a second all the data is conveyed and stored in memory in the VTR control circuit; for confirmation purposes it's decoded and displayed on the front panel of the machine or on the TV/monitor screen. When the control section detects a coincidence between stored and real-time data it switches the machine out of standby, selects the required channel and starts the recording process. At the end of the set period it powers-down the VTR and re-enters the waiting phase until the next time coincidence is detected, and so on.

An extension of this technique is used in VideoPlus remote control handsets. Every possible combination of date, time, programme length and channel is given a unique code consisting of up to seven numbers. These are published in the programme guides in newspapers and listing magazines, and keyed in by the user rather like a telephone number. They are stored in a RAM as binary data, decoded by reference to a ROM look-up table and used to select the broadcast channel and invoke the recording as described above: again the time/date/channel settings can be displayed for check and verification by the user. The advantage of this VideoPlus system, designed and patented by Gemstar, is its extreme simplicity of operation from the point of view of the user.

VideoPlus handsets are also available in stand-alone form, designed to work with any VTR which features infra-red control. In this type all the data is held in memory within the handset, which is left within 'sight' of the VTR's IR receiving eye. At the appointed moment, the handset radiates a series of codes: switch out of standby; select channel *n*; record. At the end of the pre-set period it sends the stop and then standby codes. To achieve compatibility with a wide range of VTR makes and models, the unit contains a ROM, programmed with a very large number of product codes. Each code is tried in turn by the user, who 'locks-in' the correct code when it's found: thereafter only that product code is used, and all the commands given conform to it. The same system is employed in 'universal' (replacement) handsets, which are capable of controlling four or more separate pieces of equipment once their codes have been located and logged.

Micro interfacing

The number and complexity of interfacing devices depend on the degree of dedication of the microprocessor. Late machines use purpose-designed micros which are relatively simple in their interfacing requirements; even 'analogue' outputs can be produced by a variable mark/space ratio pulse. A general block diagram for microprocessor control of a VTR is given in Figure 22.7.

Summary

In this chapter we have examined the role and operation of the VTR's system control section, and explored the interfaces by which it communicates with the 'outside world' of keyboard, tape deck and signal circuits. The evolution of syscon design has been traced from the simple logic and mechanical deck control of the early 'clunk and twang' machines to the computer-based technology of modern full-specification models, though it has not been possible to treat the latter in any great depth.

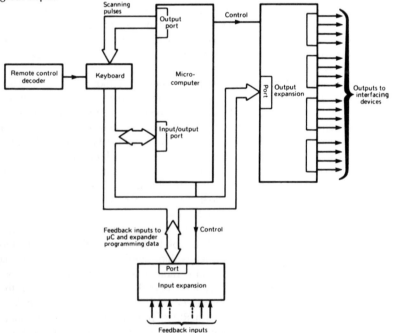

Figure 22.7 *Microprocessor-based syscon for 'housekeeping' within a VTR*

23 *The complete VTR*

The circuits and processes discussed earlier in this book form the basics of a home VTR and are, for the most part, unique (in the domestic environment) to videotape machines. In this chapter we shall deal with the peripheral circuits, i.e. those which take no part in the basic video recording process but are essential to 'service' and power the machine and provide operating convenience and flexibility. All the circuits to be discussed have counterparts in other domestic equipment such as clock-radios, TV games, audio tape recorders and TV receivers. Figure 23.1 shows a block diagram of a VTR.

Mixer/booster

To avoid regular plugging and unplugging operations, the VTR is permanently connected 'in series' with the aerial lead to the TV receiver. When the machine is off, or recording a programme other than the one being viewed, it is important that normal TV reception is not affected by the machine's presence, so a *loop through* facility is provided in the aerial booster, a small RF amplifier which is permanently powered. Its modest gain cancels the losses incurred in the extra RF plugs, sockets and internal splitting of the RF signal within the VTR.

UHF tuner and IF amplifier

Most recordings made on a home VTR come via broadcast transmissions, so the machine needs a tuner and receiver built in to select and demodulate broadcast programmes. At the time of writing, all terrestrial UK programmes are radiated in the UHF band, and the tuner and IF arrangements are identical to those provided in TV sets. An effective AGC circuit is provided to ensure a constant signal level to the recording section, and AFC feedback maintains correct RF tuning. Some VTRs have auto-set routines, in which they tune themselves, working from a program in the control section, and set their own clock from Teletext. Also found in some VTRs are built-in satellite tuners and/or surround-sound decoders.

Figure 23.1 *Block diagram of a VTR, with emphasis on the 'peripheral' circuits. The RF modulator, though marked for playback only, is often used during monitoring of recordings*

Sound

An intercarrier IF amplifier and detector follows the vision detector to provide a sound signal for recording on a tape. After the detector, the sound recording processes follow audio cassette practice, with the audio signal being added to an HF bias source before recording on a longitudinal track on one edge of the tape (as shown in Figure 17.5). The slow speed of the tape and narrow audio tape track do not make for ideal recording conditions, and noise reduction systems are commonly used to improve performance. Foremost among these is the Dolby system, which involves a form of non-linear and frequency-dependent pre-emphasis during the recording process, and complementary de-emphasis in replay.

The first stereo VTRs for domestic use used the 'split' edge-track illustrated in Figure 23.2. Currently stereo machines use one of the hi-fi systems, VHS hi-fi, AFM-stereo or PCM, described in Chapter 20. Virtually all hi-fi stereo VTRs have built-in Nicam decoders of the type described in Chapter 12. Many pre-recorded cassettes offered for sale or rent also incorporate good stereo and surround sound tracks.

Figure 23.2 *The sound track split into two for stereo sound (VHS system)*

The concept of stereo-with-TV, be it from broadcast or video tape, is a difficult one. The whole reason-for-being of a stereo setup is to recreate a wide, vibrant and 'living' sound stage in the listening area. Even with a very large screen by current standards the picture size (and position, very often) has little correlation with the sound field in most domestic situations. Stereo TV sets are now commonplace, and while they are much better than early monaural models, the loudspeaker spacing is necessarily closer, drawing in and tightening up the sound stage to match the small picture 'window'. This is not ideal for concerts and recitals, whose main feature is the sound itself! Ideally, link a hi-fi VTR to hi-fi amplifiers and loudspeakers.

Bias oscillator

To provide a suitable AC recording bias for the 'longitudinal' audio signal, and to generate an erasing signal to wipe out video tracks, a power oscillator is used.

Usually consisting of a discrete transistor oscillator built round a feedback/driver transformer, it operates at about 60 kHz, and is used only in record mode. Besides the audio head it powers a full-width erase head to wipe all signals, sound, picture, and control track, off the tape on its journey towards the head drum during recording. As described in earlier chapters, no recording bias signal is required for the video heads as this function is performed for luminance and chrominance by the constant-level FM luminance recording signal.

Clock-timer

Depending on vintage and price, VTRs vary greatly in their timing and programming facilities. All machines have an on-screen or front-panel clock display, fluorescent digital in mains-powered static machines and LCD (reflective) digital in battery-operated portables to conserve power. The clock is arranged to show real time as a permanent display, counted down from a local accurate reference – usually a quartz crystal, sometimes 50 Hz mains frequency. A digital memory in the clock/timer chip is capable of storing the clock display data for pre-set time, and this stored data is continually compared with the real-time display. When the two correspond exactly, a coincidence-detector opens a gate to invoke record mode in just the same way as a clock-radio, or clock-cassette/radio.

In the first machines, that was the limit of operation of the clock/timer, except that in some models the clock display doubled as tape counter indicator when this function is slected by the user. Modern designs of VTR have comprehensive timer facilities, typically offering a choice of eight or more pre-programmed recording 'events' of any desired length, on any channel over many consecutive unattended days. This is achieved by an extension of the above technique, with the clock detector now looking for a day-coincidence as well as time-display coincidence, then gating on the broadcast channel held in its memory as well as record mode for the tape deck. These are then held on for the requested time. Such a sophisticated timer circuit is often based on a microprocessor (see Chapter 22) which is capable of informing the operator if he keys in impossible requests like 'start at 1930 hrs, stop at 1900 hrs, or overlapped recording instructions on different channels!

RF modulator

When the machine is playing back, a TV signal at baseband is produced, and this is difficult to apply to an older type TV receiver, whose only signal input facility is via its aerial socket. To cater for this, an RF modulator is provided within the VTR, working as a tiny TV transmitter. Baseband audio and video signals are applied to its input for modulation on to a UHF carrier, whose frequency is chosen to fall into a gap in the broadcast spectrum around channel 36. A rear-access preset control swings this frequency by about ±4 channels to avoid beat effects with other RF signals in the vicinity. Some VTR modulators are programmable for any channel in the broadcast band. Modulator characteristics are according to CCIR system I, the UK standard, and a diagram of the modulator is given in Figure 23.3. Output

Figure 23.3 *Basics of UHF RF modulator*

level is set to be 1–3 mV (the optimum for a TV receiver), and this modulated RF carrier signal is added to the booster output for application to the VTR's RF output socket.

During record, regardless of the signal source which may be off-air, TV camera, cable, another VTR or whatever, the signal being recorded on tape is applied to the RF modulator so that it may be monitored on the TV set if desired. This is called the E-E (electronics to electronics) mode to distinguish it from off-tape playback mode. The E-E signal is taken off from the record electronics as late as possible, and several of the record and playback circuits (such as the luminance/chrominance adding stage) are usually included in the loop.

Set channel facility

During initial setting up, when neither the TV nor the VTR are correctly tuned, it is difficult to establish the correct tuning point for the TV when adjusting it to the output channel of the VTR's RF modulator, unless a pre-recorded tape is available to provide a playback signal. To assist with this, most machines have a simple video pattern generator incorporated, which modulates the RF carrier with a 'marker' to enable the TV to be tuned. It is often part of the multi-position colour-killer switch mentioned in Chapter 15, and is mounted on the rear of the VTR. Some VTRs generate a plain coloured raster or a pattern/text for this purpose.

Audio/video in/out connectors

Many of the signals required to be recorded are baseband for video and audio, typically coming from a TV camera or from another VTR for copying/editing. Sockets are provided in the form of (e.g.) phono for sound, DIN, BNC or phono for vision, and a 21-pin SCART socket for both.

Baseband outputs are also provided in similar connectors to those mentioned above, so that connection to audio systems, TV *monitors* and other equipment is possible. The standard for baseband video signals is universal at 1V peak-to-peak (sync tip to peak white), and for sound a level of −8 dB, corresponding to 310 mV RMS.

Record-playback switching

In almost every department of the VTR, from sound to servos, from tuner to syscon, function switching is required between record and playback modes. A multi-way slide switch to achieve this would be physically very clumsy and difficult to arrange. In fact, the record/playback switching is carried out by electronic switches in the form of diodes, transistors and ICs, toggled by supply lines which appear during the relevant mode. Thus we find '+12 V record' or '+9 V play' lines distributed to all operational blocks of the VTR. While the source of these lines is the PSU, they are enabled by the user's operating keys, usually via the syscon.

Power supplies

A home VTR consists, so far as its electrical circuits are concerned, of a lot of transistor/IC electronic circuits, a few relays and solenoids and two to five low-voltage DC motors. Most VTR circuitry operates at a supply-line voltage of between +5 and +20 V (typically +12 V), with some critical lines requiring close stabilisation and decoupling. Solenoids call more for power than precision in their supply lines, so unstabilised PSU outputs at low impedance are provided for them. Mains-operated PSUs, then, usually start with a mains transformer having several low-voltage secondary windings, each feeding a bridge rectifier followed by smoothing networks and voltage stabilisers. A typical PSU design is shown in Figure 23.4. The design is conventional and straightforward. X3/X4 form a DC series stabiliser in a control loop circuit embracing X1 and X2. The relay closes when the timer 'perks-up' and enables the main +12 V and +18 V lines. An 'ever-12 V' line for clock/timer and mix/booster is furnished by zener D6.

In industrial and TV applications, switch-mode power-supply units (SMPSU) have been common for many years; they have the advantages of higher efficiency and lower heat-dissipation over conventional circuits. An example of SMPSU design for domestic VCRs appears in Figure 23.5. The purpose-designed IC generates a squarewave drive for the power switching transistors Q1 and Q2. A square wave is developed in the primary winding of T102, whose secondaries feed a low voltage supply to a full-wave rectifier and smoothing system. Stabilisation is achieved by a 'potted-down' sample of the 12 V output which operates to vary the mark-space ratio of the square wave generated within the chip. A soft-start feature is provided by C211 and Q3, which is fed at its base with an 'enable' line from the clock/timer – Q3 thus takes the place of the relay in Figure 23.4. Over-current and over-voltage protection is incorporated in the chip.

Figure 23.4 *Simple power supply as used in early VTR machines (JVC)*

339

Figure 23.5 *Switch-mode power supply as used in a Sony VTR. The circuit is designed round a special chip, TL494CN, and offers high efficiency and close stabilisation of the two upper 12 V lines. Mains isolation takes place in transformers T102, T103 and T104. R1 is a current-monitoring resistor*

The switching transistors Q1 and Q2 are, at a given instant, either hard on or turned off, and in neither state do they dissipate much energy. This, and the relatively high frequency at which the circuit operates, leads to high efficiency, reliability and good regulation of output voltage.

Some VTR circuits require negative supply lines at low energy, a typical example being the fluorescent clock display, which calls for an operating potential of about –30 V and a filament supply of 4 V AC. These are inconvenient to generate and distribute from the main power supply, and a *DC-DC converter* is often employed, conveniently sited at the point where its output is needed, and operating from the common VTR +10 V or +12 V supply. It is a mini-switch-mode power supply, working to a similar principle as the large version in Figure 23.5, but in a package the size of a postage stamp!

Clock back-up

When the VTR is operating or in standby mode, power failure will stop its clock; this is irritating in real time, and disastrous where the machine has been primed for unattended recording. To prevent this, a small rechargeable on-board battery, usually Ni-Cad type, is provided to maintain power to the clock oscillator and counters, so that when power is restored the display lights up to the correct time. Where *volatile* memories are used, the back-up battery must also power these to prevent loss of stored data.

Non-volatile memory chips are now used in VTR equipment, and these are able to retain stored data relevant to TV channel selection, deck status and timer instructions for a long period with no power supply of any sort.

24 *Camcorders and home video*

Camcorders (camera-recorders) are the most advanced form of 'domestic' technology. Combining the elements of camera, encoder, VTR electronics, deck- and lens-mechanics and computer control, they draw on the systems, technologies and techniques described in most of the chapters of this book. Here we shall draw together these many threads and see (necessarily in broad terms) how they go together to make up the most expensive – in terms of pounds sterling per kilogram – bunch of electronics and mechanics available on the shop shelves.

Optical section

The lens assembly is a complex ensemble of precision glassware, with adjustment for focus and zoom by sliding members within the barrel. Both focus and zoom rings are driven by miniature electric motors; where manual (hands-on lens ring) adjustment is provided for these, the motors drive through slipping clutches. Zoom motor control is provided for in a body-mounted rocker switch, sometimes offering two zoom speeds. The auto-focus system consists of a servo loop with TTL (through-the-lens) picture-sharpness sensors. It's controlled by a microprocessor, and many models offer a choice of picture zones for auto-focus operation.

Also inside the lens assembly is a multi-bladed iris, whose operation is similar to that of a moving-coil meter. It too is part of a servo loop, this time controlled by the level of luminance signal coming from the image-sensor. The higher the light level in the televised scene the smaller the iris opening, with a consequential improvement in depth of focus field. There is also an AGC system in the video amplifier: between these two control loops the signal level – and hence contrast in the reproduced picture – is held constant over a huge range of ambient scene brightness. In very low light situations, however, noise (grain, snow, confetti) intrudes on the picture.

Interposed between the lens and the image sensor are two optical filters. An infra-red cut filter prevents most infra-red radiation reaching the sensor so that heat and similar energy sources have little effect on the picture, while a crystal filter takes out the finest detail in the incoming scene to prevent 'beat' and patterning effects due to the dot-matrix structure of the sensor IC and its colour filter. It also minimises cross-colour effects due to (e.g.) PAL signal bandsharing.

The image sensor itself consists of a mosaic of up to 500 000 photodiodes, each of which acquires and stores a charge proportional to the intensity of light falling upon it. As described on page 6 the charges are transferred by horizontal and vertical CCDs (charge-coupled devices) to the image-sensor's output terminal, whence they emerge (after processing in a sample-and-hold stage) as a serial video signal. Thus the scanning process is controlled in a series of small discrete steps, one pixel at a time, by the CCD driver stage in box A of Figure 24.1. Its timing is governed by the pulse generator, whose 'heartbeat' is provided by an SSG (sync and subcarrier generator) working from a precision crystal reference. As well as timing the scan and subcarrier frequencies and the colour burst and blanking intervals to broadcast TV specification, the SSG also synchronises such functions as internal caption and symbol generation for viewfinder indications and timecode generation, the latter used in precision editing. The colour components of the image are captured by a coloured translucent dot-matrix overlay bonded to the face of the pick-up sensor chip; the principle of operation is broadly similar to the striped-faceplate camera tube described in Chapter 3. Here each coloured dot-filter is aligned with its own silicon photodiode element below.

Most camcorders offer a *fast-shutter* feature, in which a very short exposure is taken in each 20 ms field period. It prevents blurring of the image when freeze-frame mode is selected during replay of fast-moving images recorded on tape. It's achieved by sweeping all the charges out of the photodiode imager at a late stage in the field period, then allowing a relatively short time-slot for integration of the image before the charges are transferred into the 'bucket-brigade' CCD.

Image stabilisation

A problem with very small camcorders *(palmcorders)* is shake and wobble in the reproduced image due to the natural tremor in the human hand which holds it. It's especially troublesome at extreme zoom settings. To mitigate this, two very different technologies have been developed. The first, brought to the market by Sony, involves a vari-angle refractive prism, consisting of two silicon-oil filled plate-glass panels linked by bellows, mounted in the lens assembly. The prism's angle, relative to the pick-up sensor's image plane, is varied by drive coils under the control of a piezo-electric sensor of pitch and yaw of the camcorder's body.

The second anti-shake system is electronic in operation. It depends on the use of a D/A converter like that described in Chapter 11 in conjunction with a large DRAM digital memory in which one whole field of picture information is temporarily stored. The image for recording on tape is read from the centre section of the memory bank, using approximately 85% of its contents, area-wise. The pixel data corresponding to the outer periphery of the picture is selectively used by the EIS (electronic image stabiliser) processor with reference to four motion detection zones in the picture and an algorithm which distinguishes between camera shake and natural movement in the picture.

Figure 24.1 *Simplified block diagram of the electronic sections of a camcorder. The lettered blocks are described individually in the text*

Digital picture zoom

In addition to the optical zoom facility provided on all camcorders, those fitted with a digital field store are able to offer a further 'zoom' feature in which only the memory-data corresponding to the central section of the picture is read out, at normal field and line scanning rates. The effect on screen is of fewer but larger pixels rather than real image magnification, so it counts more as an 'effect' than as a true extension to the zoom range. Many other effects are possible when a field store is used in a picture-processing system, as we saw in Chapter 11, and will further explore towards the end of this chapter.

Luminance processing

Returning to the block diagram of Figure 24.1, and looking now at the processes shown in section B, the iris- and AGC-regulated luminance signal undergoes *aperture correction*, in which the sharp edges and outlines of picture objects are enhanced by adding overshoot and 'undershoot' before and after the transition in the video waveform as shown in Figure 18.17. While this process does not actually increase definition, it subjectively sharpens the reproduced image as viewed on screen. In fact these contour enhancers are used on replay for the same effect; some camcorders are fitted with an 'edit' switch to soften the action of the enhancer when the camcorder tape is to be copied onto another tape in a machine incorporating its own record-signal sharpener: the effect of two such circuits acting in series can lead to spurious effects and instability.

Another function of the camera luminance processor, not shown in Figure 24.1, is that of gamma correction, described at the end of Chapter 2. It's achieved by a voltage-controlled attenuator (VCA) circuit in which the control voltage is derived from the input signal itself. The way in which the luminance gamut is stretched is shown in Figure 24.2. The luminance signal has blanking, sync pulses and colour bursts added at the mix box on the right of block B in the diagram. At this point, too, the chroma subcarrier signal is superimposed, and we shall next examine its derivation.

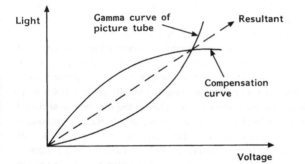

Figure 24.2   *Gamma-correction curves for light output compensation in a cathode-ray picture tube*

Chrominance processing

Box C in Figure 24.1 is deceptively simple! In practice it incorporates a great deal of IC-based processing circuitry, which in advanced models is carried out (along with the luminance signal processing) in the digital realm.

The first step is to derive R and B primary-colour signals from the CCD image sensor output. It's done by a pair of sample-and-hold gates driven by a pulse output from the CCD driver circuit. The pulses are timed to coincide with the repetition frequency of complementary colour filter dots on the sensor faceplate.

The samples thus obtained are fed to a switched matrix whose outputs form R and B signals with relatively low definition due to the 'coarse' nature of the optical colour filtering system. This is not a problem because the U and V signals derived from them have limited bandwidth anyway, as we saw in Chapter 3.

At this point the white-balance correction is applied. Different light sources (daylight, sunlight, artificial illumination etc.) have widely different hues. Human eyes can 'neutralise' these when viewing the actual scene but not, strangely enough, when they are reproduced on screen. All must be rounded up or down then, colour-temperature-wise, to the standard Illuminant D (see Chapter 2) for television purposes. Some camcorders have pre-selectable settings for white balance, typically correcting for indoor/outdoor conditions. All have an automatic facility for white balance (AWB) in which the incoming light is sampled, either through the lens or via a separate light-sensor on the camera body, and the gains of R and B amplifiers set accordingly. If the scene being viewed is a white card – or simply the ambient light via a translucent white lens-cap – the colour-balance is correct when both the R – Y and B – Y signals are at zero. The gain settings for R and B amplifiers can be stored in memory at each sampling and used thereafter until different lighting conditions obtain, when once again the sample-and-store process is manually invoked. Some camcorders continually auto-adjust colour balance on the basis of the sampled hue of the picture, but this can lead to desaturation and other spurious effects on certain types of scene.

Only when the colour balance is spot-on will white objects in the televised scene appear truly white in the TV picture, and only then will other hues be correctly reproduced. The most critical ones in this respect are flesh-tones, one of the few 'real' references for the viewer of a TV picture.

Now the B–Y and R–Y colour difference signals are derived in a matrix involving a bandwidth-limited luminance (called Y_L) signal from block A in Figure 24.1. They are weighted to become U and V components and encoded into a PAL (or whatever) chrominance signal as described at length in Chapters 3 and 4 of this book. The chroma subcarrier joins the Y signal in the mix box of section B of the diagram. In high-band (S-VHS, Hi-8) camcorders the Y and C components are kept separate throughout to avoid the defects (primarily cross-colour, see the end of Chapter 4) of band-sharing inherent in PAL (Secam, NTSC) broadcast transmissions. In our camcorder example, however, the video signal is a composite one, now ready to leave the camera section. It takes three paths: to the tape-recording electronics, to the video output socket (for monitoring) and to the viewfinder, which we'll examine next.

EVF

The electronic viewfinder (EVF) can take several forms, but most often is based on a minature black and white picture tube with a very small deflection angle and a high-definition screen, supported by a tiny bunch of electronics in the form of a single IC and a few peripheral components. These latter include a small line output section – transistor and flyback transformer – to provide scanning current in the line deflection yoke, and operating voltage for the electrodes of the mini-

CRT, as shown in section D of Figure 24.1. The whole ensemble, as can be seen in Figures 26.1c and d is barely bigger than half a sausage even though it contains all the elements of a monochrome monitor, the operation of which is discussed variously in Chapters 1 and 8 of this book. The tiny screen is fronted by a slide- or twist-to-focus magnifying lens built into an eyecap.

Peripheral to the EVF, and drawn below it in our diagram, is the character- and symbol-generator which superimposes data on the EVF screen. The readouts thus provided can give the operator a wide variety of indications and prompts: battery charge state, light level, record/pause status, tape counter, focus zone, date/time (this can if desired, be superimposed on the recorded picture, too) white balance status, etc. etc. These are gathered by sensors associated with the control system and composed by a (usually bus-controlled) caption generator IC in synchronism with the field and line scan timings.

Alternative EVF systems include colour LCD screens of the sort described in Chapter 7, some types of which occupy much of the rear surface of the camcorder case, removing the need for an eyecup magnifier system. They have the advantage of giving the operator some idea (though not as accurate a one as a conventional colour monitor) of the colour-balance of the pictures being recorded. The second defect of colour LCD EVFs is that they do not have the definition capability of a tube-based system and so do not facilitate such accurate focusing.

Both of these problems are neatly solved by the colour-wheel EVF display system whose principle is shown in Figure 24.3. Here a black and white picture tube is used in conjunction with a three-section rotating colour-filter wheel. The mini-cathode ray tube is fed, in rapid succession, with R,G and B video signals. Filter presentation, in RGB order, is synchronised with these by a phase-control system working on the wheel-drive motor, similar in essence to that used – as described in Chapter 21 – to control the positional phase of the video head-drum. This type of field-sequential colour-display scheme was one of the first to be proposed and tried for colour television many decades ago!

Figure 24.3 *Principle of sequential colour display system as used in some camcorder viewfinders*

Audio section

All camcorders have an integral microphone, which may be a mono or – with hi-fi camcorders – a stereo type. Many have external microphone sockets for greater versatility and an opportunity to eliminate the motor- and handling noise which is almost inevitable in quiet situations where the ALC (automatic level control) drives up the gain of the audio amplifier. The action of the ALC circuit is necessary to enable the camcorder to cope with the huge range of sound levels it may encounter, from the rustle of leaves to the roar of a jet aircraft. All these have to be brought within the dynamic range of the tape system, discussed in Chapter 16. Where a hi-fi sound recording system is provided in the more expensive models, the dynamic range for sound recording is much greater and the ALC action can be less harsh. The camera sound is passed out of the camcorder during record along with the composite video signal for monitoring if required: very often a headphone socket is also provided for audio monitoring 'on the hoof'.

Sound recording and playback techniques were discussed in Chapter 20, and the processing circuits of camcorders conform to them. In VHS camcorders using small head-drums (Chapter 17) the provision of the extra heads for depth-multi-plex recording makes for a complex and expensive drum assembly on the little tape deck! Video-8 and Hi-8 formats use AFM or (top-grade models only) PCM sound systems, again like those covered in Chapter 20.

Camcorder tape deck

Section E of Figure 24.1 depicts the essential parts of the tape deck, whose main distinction, compared with a homebase deck, is its extreme miniaturisation. Small head drums are used, with the multi-head and long-wrap techniques described in Chapter 17. The servos are digital types developed from the basic concept illustrated in Figure 21.17. The capstan and cylinder drive motors are physically very flat with direct drive (Figure 21.12) and no electrical moving parts, and are designed to react fast enough to counter the effect of camera movement.

The electronic (sound and vision) processing circuits associated with the tape deck are the same, in essence, as those already covered in the relevant chapters of this book; again the main difference is in the physical size of the components and circuit boards. In camcorders wide use is made of sub-miniature and surface-mounted components, mounted on tiny double-sided or multi-layer printed circuit boards, some of which are flexible, and most of which are peculiar in shape to match the profile of the case and the mechanical tape deck. An idea of the internal construction of the camcorder is given by the photo of Figure 24.4.

Control system

In a camcorder the control microprocessor has many more interdependent functions to control and co-ordinate than in a homedeck VTR, although the basis of its operation is similar to those already described for TV sets and VTRs. Although the

Figure 24.4 *Inside a Sony camcorder: (a) optical block and picture processing sections; (b) circuit boards unfolded to show layout and flexible inter-board connectors*

system control heart in Figure 24.1 is shown as being based on – and confined to – the mechanical deck section, its electronic tentacles extend over every section of the camcorder, from the lens drive system to the AV output switching and power supply section. Control is effected, except in early models, by one or more two-wire serial bus communication systems, which also interface with subsidiary microprocessors such as that which controls auto-focusing.

An EEPROM is linked to, and governed by, the control microprocessor in similar style to that described in connection with the TV control system outlined in Chapter 8. Here its purpose is not to store broadcast-reception tuning points; it holds data on set-up levels and reference-points throughout the camcorder, set by the user, the service engineer and the manufacturer. It contains such diverse data as the head-switch point, the white-balance setting, audio record level and PAL encoding parameters. No longer does a factory operator or a service engineer twiddle pre-sets in the camera section, PAL encoder or luminance circuits: the data for these is written into the EEPROM via the remote control handset or from an external computer, whence they are conveyed during camcorder operation to the appropriate circuit section and there (where relevant) translated into control voltages or currents by D/A converters within the processing ICs on the spot. This simplifies and speeds up factory production and eases the job of the service technician – so long as he has the necessary hardware, software and data to do the job! Allied to this concept are diagnostic readouts, generated by the control microprocessor, to relay to factory and service personnel fault data, typically: deck shutdown was invoked because of lack of rotation of the take-up spool/ incorrect frequency of drum tacho pulses/excessive tape-loading period – or whatever.

Some camcorders have a serial data link to the control microprocessor from an edit-control port, using an operating code similar to that of the remote control system we examined in Chapters 8 and 22. Examples of these are the Sony-designed LANC system and the 5-pin link system incorporated in Panasonic camcorders and homedecks. They permit control of the camcorder by an editing system of the sort we shall look at shortly here.

Time-code generation

Also designed for use in post-production editing, some of the more expensive camcorders incorporate a *timecode generator*, in which every TV frame recorded on tape is given its own unique electronic 'label' in terms of elapsed hours, minutes and frames. These labels can be detected by suitable editing machines and used to control a copying tape deck for precise and frame-accurate editing. There are two timecode systems in common use in home camcorders. Video-8 and Hi-8 types use RCTC (re-writable consumer time-code) in which eight-bit digital data bytes are stored on a video tape-track extension between the video and PCM (where incorporated, see Chapter 20) portions. This is shown in Figure 24.5. VHS-based camcorders incorporating timecode use a similar system called VITC (vertical interval time-code) in which the data are carried in the unused lines of the field blanking interval – the area which carries teletext data in broadcast TV transmissions. The advantage of RCTC is that it can be written (in a suitable camcorder or video deck) onto an existing tape recording, whereas adding VITC to a recording involves copying it onto a second tape in the VITC-equipped VTR.

VHS-C camcorders

The Video-8 cassette is a small one, intended specifically for use in camcorders; it permits the design of very small 'palmcorders'. The VHS system, on the other hand, is based on a much larger cassette originally designed for home-deck VTRs;

Figure 24.5 *In Video-8 format the RCTC timecodes are written onto the tape at the start points of the video tracks. Not all V-8 VTRs use the PCM sound tracks shown dotted here*

its use in a camcorder makes for a large, bulky and heavy unit. To overcome this problem the VHS-C (compact) cassette made its appearance. It contains enough tape for a maximum shooting time of 1 hour (SP) or 2 hours (LP) and forms a much smaller package. Even so, it retains compatibility with the VHS – or S-VHS – format, and can be operated in a full-size deck; in most cases an adapter, whose body-shell conforms to VHS dimensions and mechanics, is required to facilitate this. More on this and allied subjects will be found in Chapter 26.

An internal view of a camcorder is shown in Figure 24.4.

Home movie post-production

It is seldom that the tape which comes straight from the camcorder provides a complete and satisfactory 'programme' by itself. As in the days of amateur cine cameras it's usually necessary to cut out the mistakes and the unwanted footage, and to reassemble the rest in a different order to that in which it was shot, while inserting titles and perhaps adding special effects.

Dubbing and editing

Audio dubbing is carried out by operating the machine in replay so far as the vision signals are concerned, but with the sound section of the machine in record mode, and the audio input jacks 'live'. Apart from the necessary electronic switching facilities, the machine needs only an additional audio erase head (positioned on the tape path just before the audio rec/play head) to achieve this, and in dub mode the replay picture is monitored while adding a new sound track to the recording.

With cine film the picture editing process is a physical one, involving scissors and glue. For half a dozen good reasons videotape cannot be cut and spliced, so the technique is changed to that of selectively copying sections of the original footage onto a second tape in the required order, and thus building up a *master copy*, on which the soundtrack, too, can be composed and controlled by the editor. The problem here (except with digital tape recording systems) is that each time the picture and sound are copied from tape to tape their quality is compromised, especially with low-band domestic formats. It's important, then, to limit the number of 'generations' involved in the editing process, and to get the best-possible quality of picture and sound at the outset.

The basic editing process consists of setting the source VTR (generally in budget-limited home operation, the camcorder itself) to play, the copy VTR (which may be the family's general-purpose machine) to record, then to operate the latter's record-pause control to hold up the copying action as each new scene or sequence is found and set to play in the source machine. This is called *assemble editing*. Ordinarily this would lead to 'glitches' and instability in the played-back picture at the transition points due to discontinuities in the vision/sound recording and the control track. There are a number of reasons for this. During record and playback the servos, as we have seen, are locked up to incoming field syncs and off-tape control tracks respectively. When a change is made in input signal the servos have to run up to speed and lock-in to the new video source, and so the signal recorded on the tape goes haywire during this period. On replay the discontinuity in the control track also causes a 'hiccup' in servo operation, and these effects result in a very ragged transition between the two pre-recorded sequences. Further problems arise from the fact that the full-width erase head is placed about 8 cm before the video head drum on the tape path, giving rise to three or four seconds crosstalk from the original tape tracks before newly-erased tape reaches the recording heads.

'Clever' edit

To overcome these problems and give a reasonably clean transition at edit points, the technique of *back space edit* was introduced. In its most common form, Figure 24.6, it involves the VTR machine back-spacing (or rewinding) 20 to 25 frames (about a second) when record pause is selected, then stopping. At the end of the pause period the machine will re-start in playback mode so far as its vision systems are concerned, but with the capstan servo slaved to the new incoming video signal. When the servo has locked up, the vision circuits switch automatically to record mode during a convenient field-blanking interval so that the last frame of the old material is followed by a new and fully synchronised field of the new programme with no lack of continuity in control track or video tape tracks; very little disturbance will be seen on the replayed programme over the edit point. A certain amount of chroma crosstalk can occur briefly at the changeover point in some designs, and much depends on the sophistication of the VTR's electronics – this is another area in which the microprocessor has come to the fore.

The alternative and less-used *insert-edit* system consists of the 'dropping-in' of a sequence into a pre-recorded programme, examples being titles, credits and commercials. Here the original control track is retained to give 'glitch-free' transitions at the start and end of the insert. Some home-decks and camcorders are equipped with ports for *synchro-edit*, in which the pause and run functions of both VTRs can be controlled by a single key at one end or the other.

Manual editing is a long and wearisome business, especially when the required sequences are scattered widely along the source tape and the programme being assembled is a long one. It can be automated to a greater or lesser degreee by an

Figure 24.6 *Assemble edit: the sequence of events is indicated by the ringed numbers, and the result is a clean switch from old to new programme material*

edit controller, which may be incorporated into the source or copy machines; provided as a stand-alone unit; or take the form of a home computer working in conjunction with suitable software and interfacing hardware. All work on the same basic principle of storing the (human) editor's cut-point instructions (known as an EDL, edit decision list) in a memory and manipulating the transport controls of source and copy VTRs to assemble the master copy as required. In the simplest systems the cut-point references are defined by an electronic frame-counting process. This becomes less accurate as time (and frames!) passes, and more sophisticated – and expensive – systems make use of the timecode schemes mentioned earlier in this chapter.

One of the simplest auto-edit outfits is based on the random assemble (RAS) concept, introduced by JVC, and incorporated in the camcorder illustrated in Figure 26.1c. Here the edit memory is built into the camcorder, and programmed by the user keying in the required start and finish points of each sequence required. A hard-wire link is provided for connection to a suitably-equipped homedeck, and when set to go in this mode the camcorder finds and runs the selected sections of tape, controlling the pause function of the copy deck meanwhile.

Stand-alone and computer-based systems work on the same principle, and have provision for storage of up to 200 'edit events'. For them to work there has to be some form of access to the camcorder's (or source machine's) control system so that the deck can be manipulated. Control of the copy VTR can be achieved by a hard-wire link (e.g. LANC) or by an infra-red wand whose emitted signal mimics that of the deck's own infra-red remote control unit. Some edit-control systems are capable of operating by frame-counting, VITC or RCTC, and of controlling the slave VTRs by LANC, Panasonic, pause-jack or infra-red means, offering great versatility in operation.

Computer-based edit systems, as well as governing the automatic-assembly modes described above, can also generate titles, captions and effects and (with the help of a *genlock*) synchronise them with the video picture and superimpose them upon it. The basic setup for harnessing a computer to home video production is shown in Figure 24.7. Packages are available for all the popular types of home computer, and some of them can offer a wide range of facilities including frame-grabbing, animation, digital data-rate compression and audio processing.

Figure 24.7 *Use of a genlock, computer and sound mixer to enhance an edited video tape*

Simpler units available for home editing include enhancers, which artificially sharpen up images in the way shown in Figure 18.17; colour correctors, which decode the composite signal to Y and RGB to facilitate control of each colour individually for hue manipulation, then recode to (e.g.) PAL standard; caption generators containing a keyboard and picture-synchronised character-generation IC system; video inverters for viewing photographic negatives and for effects; and sound mixers in which a variety of audio sources can be blended, faded and mixed for recording on the master tape.

Digital effects units

By converting complete video fields into digital form and storing them in a large memory as described in Chapter 11, it becomes possible to achieve the effect of genlock, whereby two autonomously-generated pictures can be mixed, superimposed or cross-faded. One of the video signals is A – D converted and written into field memory in seven-bit or eight-bit form, to be read out again, one or two fields later, at a rate governed by the synchronising pulses of the second picture source as illustrated in Figure 24.8. In addition to the basic 'genlock' function, many

Figure 24.8 *Digital frame synchroniser for superimposition of TV images from two autonomous sources*

Figure 24.9 *Picture-in-picture is one of the many effects afforded by a digital picture field store*

other effects become possible when a one- or two-field picture memory is available: 'wipes', mosaic effects, picture zoom, still-frame and strobe images, picture-in-picture (Figure 24.9), and chroma- and luma-key, in which selected types of background in one picture can be replaced by the corresponding parts of a second picture. Thus, for instance, a person shot against a plain blue background can be made, on the master-tape or the monitor screen, to appear on a beach, in a city street or on the railway track with a fast-moving train approaching!

Timebase correction

As we have seen earlier in this book, the necessity for videotape replay to depend on a mechanical process imparts to the reproduced video signal a certain amount of timing jitter. In multi-generation video copies (the result of editing processes) this jitter can be sufficient to upset the line synchronisation of the TV or monitor used for display; the result is horizontal instability of vertical edges in the picture, and often a bending or wobbling at the top of the image, perhaps with hue errors there as a result of poor synchronisation of the colour subcarrier regenerator in the chroma decoder. This problem can be overcome by the use of a timebase corrector (TBC) whose principle of operation is the same as that shown in Figure 24.8. The jittering video signal is A/D converted and the data produced written into memory one field at a time. Memory readout is governed by a stable local crystal oscillator which also times the newly-generated outgoing sync pulses. Indeed the sort of effects unit shown in Figure 24.7 has a TBC function so long as the video-2 input (master) has no timing jitter superimposed on it. Some top-class camcorders and home VTRs/edit decks have digital TBCs built in – they are switched into operation during the replay mode.

25 Care, operation and maintenance of VTRs

A videocassette recorder is a complex ensemble of mechanical, electrical and electronic parts working to very high precision. As with any such equipment, careful handling and regular routine maintenance are necessary to maintain performance throughout the life of the machine.

Operation of the domestic VTR is relatively simple with the exception of some clock-programming procedures, when the instruction book must be slavishly followed! Much thought has been given by designers to ease of operation, and the cassette system and syscon between them make the machine virtually foolproof. The VTR should be installed on a strong, level surface in a position where the ambient temperature is reasonably constant and the atmosphere clean – the 'chimney corner' is definitely out on both these counts! It is recommended to buy a dust cover – regular use of this will prevent ingress of atmospheric dust and soot particles, so prolonging the machine's life. In high ambient light, and particularly direct sunlight, the usable range of any infra-red remote control system will be greatly reduced, and this should be borne in mind during installation.

If the VTR is in close proximity to the TV receiver or monitor, mutual interference effects can take place due to direct radiation of signals between them. This leads to spurious effects ranging from patterning and striations to complete or partial loss of colour. While the specific cause of the effect may be obscure, physical separation of the TV and VTR will prove the point, and these symptoms can often be cured by installing a metal or foil screen between the two equipments, and earthing it if necessary.

Tape care

Considering its nature, video tape is surprisingly uncritical in storage and handling, largely due to the protective cassette housing. Cassettes should not be placed near magnetic field sources such as TV sets, large loudspeakers or transformers; they should be stored standing upright to prevent sag or cinching, preferably in an even temperature away from dust.

When a tape is used for 'time-shift' or similar purposes, the temptation is to rewind it to the beginning before each recording so as to have maximum recording time available. This practice leads to disproportional wear on the first part of the tape, resulting eventually in noisy recordings with lots of drop-out. Very often a three-hour tape is used for general purposes and the first 30 minutes-worth can become worn before the tail-end of the tape has seen any action at all! It is far better to work through all the tape, using the counter memory, before starting again at the beginning.

Head-cleaning tapes

The use of proprietory cleaning tape cassettes for domestic VTR head-cleaning is something of an emotive issue, with contradictory claims being made for their effectiveness, and much more important, the possibility of their causing damage to the video heads, by their makers on the one hand, and VTR manufacturers and the service trade on the other. If a head-cleaning tape must be used, the soundest advice is to use a type marketed or recommended by the VTR manufacturer. Leaving the head-damage claims aside, however, there is no doubt that hand-cleaning of the heads (and the rest of the deck) is far more effective than any cleaning tape, although the former course is not open to the average VTR user, except by arranging service from a specialist service organisation.

Deck servicing

It must be emphasised here that any attempt to clean tape heads or adjust deck components by a person without the necessary knowledge and equipment is almost bound to end in disaster! One half-turn on a tape-guide adjustment screw can destroy the machine's compatibility, and the lightest of cleaning strokes in the wrong direction across a video head will destroy it; the repair bill could run to three figures. VTRs are not like the car or the central heating system!

Deck servicing resolves into three main areas, those of cleaning heads, tape path and deck surfaces; adjustment of tape guides and tensions to ensure compatibility; and attention to secondary mechanical parts such as brakes, clutches, drive wheels and operating levers and sliders. We'll look at each of these aspects in turn.

Cleaning

In Chapter 17 we studied the requirements of compatibility and saw that this calls for great accuracy in the laying-down and reading-off of video tape tracks. This is governed entirely by tape guides and head-drum ruler edge, and the angles and dimensions of these components are critical. After a period of use, dirt and grease deposits build up on their surfaces, and that of the head drum. These can obstruct or divert the tape path, and the result will be mistracking. The first step in deck

Figure 25.1 *Cleaning points in a typical videocassette deck. The shaded parts are the tape heads, and cleaning of these should be left till last*

servicing, then, is to thoroughly clean the parts illustrated in the typical deck layout in Figure 25.1. They are the tape guides, rollers, ruler edge, head drum surface, capstan and pressure roller. If necessary the latter should be replaced rather than refurbished. At the same time, residual dirt, tape particles and grease should be removed from all deck surfaces to prevent its later migration on to critical components. A cotton bud and methylated spirit are the best tools here, the moistened bud forming an effective wiper and collector of debris. When all surfaces are clean and guides, rollers and capstan polished, head cleaning can begin.

For the static heads (full-width erase, sound and sync) a perfectly clean cotton or buckskin cloth should be used, moistened with alcohol; a cotton bud is useful where head surfaces are difficult of access, as in some U-loading machines. Firm rubbing across the face of the heads is required, followed by final cleaning and polishing with a clean cloth or bud. The video heads are very fragile indeed, and cleaning should be undertaken with very great care. Manufacturers recommend the use of a lint-free (i.e. chamois or buckskin) cloth dipped in methanol, though many engineers prefer pure surgical spirit as it leaves less deposit. The cleaning motion should not be prolonged, and must always be made across the heads (i.e in the direction of the tape path), *never* vertically. This is illustrated in Figure 25.2. Suitable cleaning materials and solvents are available from VTR manufacturers. For inspection of heads and guides during cleaning a dental mirror is very useful, especially an illuminated type. After cleaning, it is important to allow all fluids to dry completely before tape loading, or the moist surfaces will 'pick up' the tape with dire results. A hairdrier can be used to speed up the drying process if necessary.

Figure 25.2 *The cleaning motion for video heads must be side-to-side. The video head is very delicate indeed; its width may be less than that of a human hair*

Tape guide alignment

There are several tape guides encountered by the tape on its path through the deck, as shown in Figure 17.9, and the most critical of these are the entry and exit guides at each end of the tape's wrap around the head drum. These determine the tape angle relative to the fixed video head sweep, and are made adjustable so that exact compliance with the format specification can be achieved. For correct setting-up, a standard is required, and this comes in the form of an *alignment* or *interchange* tape, pre-recorded at the factory on a precision standard machine under closely controlled conditions.

The entry guide determines head tracking angle at the point where the heads start their sweep of the tape, and examination of the RF *FM envelope* signal from the replay heads will indicate any fall-off in amplitude here due to guide misalignment. A good envelope signal is shown in Figure 25.3a, while b indicates low FM output at the LHS (start) of each envelope pattern due to incorrect entry-guide height. The exit guide is adjusted for optimum envelope shape at the end (RHS of oscillogram) of each head sweep, and c shows the effect of maladjustment here. Because of the FM limiting circuits employed on playback, a worst-case FM envelope shortfall of 40 per cent is tolerable before deterioration of the picture is noticed; this is illustrated in Figure 25.4.

The other tape guides are concerned mainly with defining the tape path across the static heads, and ensuring a parallel passage through the capstan/pressure roller combination. The adjustment of the sound/sync head is critical for height,

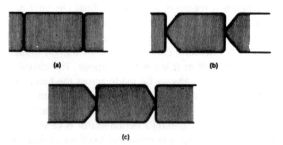

Figure 25.3 *RF envelope patterns: (a) shows a good envelope, (b) one resulting from a dirty or incorrectly adjusted entry guide, and (c) the effect of dirt or maladjustment in the exit guide*

Figure 25.4 *The limiter will clip the FM luminance signal on replay, so that a shortfall of up to 40 per cent will pass unnoticed*

azimuth and vertical tilt, and these are set up (on the audio signal from the alignment tape) by adjustment of the head's base mounting screws, in similar (but more critical) fashion to audio-recorder heads.

Tape tension is set by adjustment of the tension servo, in whatever form it takes (see Chapter 21), for a specific back-tension, monitored by a suitable gauge supplied by the manufacturer. Take-up tension is determined by the friction in the take-up clutch. This is either fixed during manufacture or is adjustable in three or four pre-set steps. VTRs with separate reel motors may have an electrical pre-set TU tension adjuster.

Reel drive, alignment and braking

The final aspect of deck servicing is concerned with the reel transport mechanics and operating levers. Their intricacy will depend on the vintage of the machine; early types require lubrication and occasional physical alignment of spindles, levers and slider bars; and roughening and degreasing of rubber friction wheels and drive surfaces. Spirits can attack rubber compounds, and a mild detergent solution is recommended for cleaning and degreasing drive belts and rubber wheels. Sometimes this is easier said than done, and replacement rather than refurbishing of interwheels, brake pads and drive belts is often the most practical course. Later machines with direct-drive systems are easier to maintain in this respect.

There are several other critical adjustment points in the video tape deck, such as reel disc height, tension pole positioning and video head eccentricity, and the setting of these is comprehensively covered in the relevant service manual for each model. Jigs and tools are available from VTR manufacturers to assist with precision adjustment, and a set of these for one particular machine is pictured in Figure 25.5.

Electrical adjustment

Many of the cautions mentioned above are also relevant to the electrical pre-set controls within the machine. Although electrical setting-up (especially of those controls which concern servo and video-head response characteristics) is a part of

Figure 25.5 *Tools and jigs required for use in VTR deck servicing. The tool kit incorporates a tension/torque-measuring cassette*

routine maintenance, electronic devices are not subject to wear; and little drift is encountered in modern circuits. Electrical adjustments tend to be specific to each model and format, so that little general guidance can be given here, except to say that service manuals give precise setting-up instructions to be used with the necessary test gear.

Test equipment

For VTR circuit investigation and servicing, it is important that the test equipment used be adequate, accurate and dependable. An oscilloscope of minimum specification would be 10 MHz bandwidth, 2 mV sensitivity and dual-trace capability. High-impedance multimeters of both digital and analogue type are required, the former with a DC voltage accuracy of 0.5 per cent or better. For investigation of colour-under and servo electronics, a frequency counter is essential, preferably a type with $6\frac{1}{2}$ or more digits and an accurate timebase.

Necessary signal sources are a bench-type colour bar signal generator with split-field display and outputs at video (baseband) and UHF; an accurate signal generator with sine- and square-wave outputs; and a source of noise-free off-air transmission signals at UHF. For monitoring purposes a modern high-performance TV receiver is ideal, preferably with a video input in monitor fashion. A mains variac is very useful for investigating power-supply and excessive-loading faults; a mains isolating transformer is also important for safety, especially when dealing with the sort of switch-mode mains power supply described in Chapter 23.

In addition to these, alignment tapes and adjustment jigs are required as described earlier in this chapter, along with a full service manual for the machine under investigation. The final factor is a good working knowledge of the principles and practice of the VTR!

Output channel and still-frame adjustment

Some VTR models have rear-mounted pre-set adjustments for use during installation, and the two most common will be described. When the VTR is being used in the environment of other RF generators around TV channel 36, the machine's RF output will beat with them, giving rise to dots or a wavy or herringbone pattern on the reproduced picture. Although no broadcast TV transmitters operate in this region, some radar installations give rise to spurious outputs at about 600 MHz which can upset VTR reproduction. Increasingly, other UHF signal sources are finding their way into the home, in the form of TV games, computers and satellite receivers, most with UHF modulators operating in the region of channel 36. To avoid clashing with these, the VTR's UHF modulator can be varied over (typically) eight channels by adjustment of a trimmer in the RF modulator. When adjusting this, go only so far as is necessary to eliminate beat effects; RF output level drops off at the extremes of adjustment, and on a more practical plane, the tuning slug may drop into the works if it's screwed in too far!

Where noise-free still-frame facilities are provided in a VTR, there may be an external adjustment for pre-setting the 'artificial' field sync pulse timing during this mode, as described in Chapter 21. In 'freeze' mode a degree of judder may occur if successive fields differ greatly in content, as would be the case when televising a fast-moving athlete. The ideal frozen frame for this adjustment, then, is a test card or similar stationary picture with well-defined horizontal edges, on which the control should be set to 'null-out' vertical judder in the displayed still frame.

Television time base requirements and modifications

In our discussion of luminance and chrominance playback techniques (Chapter 18), we came across the phenomenon of timing errors in the off-tape replay signal, known as jitter. The deck maintenance and lubrication procedures just described go a long way to minimise jitter, but it is always present in the VTR's output signal and can upset the horizontal synchronisation of the TV receiver (or monitor) unless the latter is designed for use with a VTR. TV sets are primarily intended for broadcast reception, and in noisy or fringe reception areas line synchronisation can be upset by interference and electrical noise. To overcome this problem, they are fitted with a flywheel line-synchronisation circuit in which the timing of the line sync pulses over several lines' duration is 'averaged' to effectively slow down the response of the line ocillator to them, as described in Chapter 8. Thus occasional ignition interference or noise pulses occurring just before the end of a TV line are prevented from falsely triggering the scan generator and giving a ragged appearance to the picture.

If the flywheel time constant is too long, the timing jitter on the VTR output signal (which is present, of course, on picture information as well as line syncs) will cause the picture to twitch horizontally while the timebase produces a smooth, average-rate scan timing. This effect is particularly annoying, and to prevent it the flywheel time constant must be reduced to enable the timebase to follow the rapid off-tape timing fluctuations, thus straightening up the picture. For many years TV design has taken account of this, and a dedicated VTR button on the set automatically switches in a suitable time constant in the flywheel department in sets where this is not present anyway. Early sets need modification to their flywheel circuits to fit them for good video playback, and this generally involves reducing the time-constant to about one-third of its original level – specific modification details are available from set-makers and VTR manufacturers. Some TV receivers automatically select the required time-constant.

Receiver-monitors

Most contemporary TV sets have advanced power supply circuits along the lines of Figure 8.9 which give mains-isolation to the 'ground' line of the set, and this paves the way to the provision of input sockets for audio and video signals at baseband. The TV becomes known as a receiver/monitor, and where this type of set is available, much better replay performance in vision and sound is possible by direct connection to the VTR. Baseband linking eliminates the degenerative processes of converting to and from UHF and IF, and has much to recommend it. The use of a purpose-designed precision monitor is not really justified with domestic VTR machines, as system performance is limited by the VTR capabilities, which fall far short of those of the monitor.

Distribution of VTR output signals

Where a VTR is required to operate several displays simultaneously, perhaps at widely different points, there are two choices in the mode of distribution. For optimum quality and avoidance of spurious effects due to patterning and timing errors, the ideal is to distribute at baseband. This involves a coaxial cable carrying composite video at 1V peak-to-peak to monitors or monitor/receivers. A video distribution amplifier is needed to provide outputs at 75 Ω impedance for as many monitors as required, while the sound signal is either carried in a separate cable or sent along the video coax, with insertion and extraction by suitable low-pass filters at each end. Such a system is ideal for teaching, conferences and similar applications and will operate at distances up to 700 metres without serious distortion.

The alternative is distribution by an RF carrier, and so far as domestic VTRs are concerned, this implies the use of the UHF output from the machine. If more than two monitoring points are needed, a UHF distribution amplifier will be necessary, and by this means many sets can be fed with the playback signal, at great distances if required. This method, while offering lower picture and sound quality than the baseband system, has the advantage of convenience in the use of easily-available equipment. The two systems are shown in Figure 25.6a and b.

Figure 25.6 *Two forms of signal distribution for VTR playback. That given at (a) is capable of best results*

VTR operation on non-standard signals

There is some confusion over the question of interchangeability of VTR machines and tapes between different countries of the world, probably arising from the fact that all three major formats are used world-wide, and it's reasonable to suppose that, for instance, a VHS format colour tape recording made in the USA will replay on a UK-type VHS machine. In fact all the basic formats are manufactured in export versions for different world markets, but within each format some differences are present in the electronics to cater for variations of colour encoding (PAL, SECAM, NTSC), line frequency (525 or 625 lines) and field rate (50 or 60 Hz), and these determine the circuit design of servos, colour-under systems and still-frame facilities.

The different transmission standards between countries add further complication, this time in the peripheral circuits of the VTR such as tuners, sound detectors and RF modulators. Vision and sound modulation systems differ, and VHF bands are used for TV transmission in many parts of the world. As a result, UK-type machines will only work in Eire, Hong Kong and South Africa, though the tapes are interchangeable with all countries using PAL/625/50 standards. We shall see in the next chapter the effects of incompatibility between formats rather than TV standards.

Some manufacturers market multi-standard machines which go some way towards solving the compatibility problems arising out of differing TV systems. Very few truly universal machines exist, however, and if a multi-standard machine is considered for purchase, it's important to study the specification carefully and establish that it will do what is required of it; these machines are usually only available to special order.

26 Tape formats – systems and facilities compared

In preceding chapters we have examined VTR principles and circuits in general terms, taking examples from all home formats to illustrate the techniques used. Where differences between formats have arisen in the text they have been surprisingly few. In this last chapter we shall expand on the format differences and look into their significance, both from the point of view of the consumer and those concerned with the technicalities of the machines.

There are currently four major analogue home videotape formats in the consumer field: VHS, S-VHS, Video-8 and Hi-8. VHS owes its origin to JVC, while the Video-8 formats are later developments, using more advanced features – we have met ATF, RCTC, AFM and PCM sound in previous chapters. Judged subjectively on normal programme and home-movie material, the performance of the low- and high-band variants of VHS and Video-8 respectively is quite similar in terms of picture resolution and S/N ratio.

Over the years the competing formats have shaken down into well-defined categories. Video-8 is favoured for camcorder use, while VHS is the system for general-purpose homedeck use, and virtually the only one in which pre-recorded cassettes are produced for sale and rental. The high-band variants Hi-8 and S-VHS find a relatively small market, even though they are capable of producing better pictures than the others. This suggests that the public are more concerned about price than image quality! A tribute to Hi-8 is its gradual acceptance by TV broadcasters for certain types of programme, typically ENG (electronic news-gathering) and low-budget 'community' contributions. Examples of home video equipment for the four formats are shown in Figure 26.1

The existence of four incompatible formats – and two cassette-size variations for VHS – is inconvenient for the user and sometimes exasperating for the trade, but is indicative of the atmosphere of free enterprise and unrestricted marketing which we enjoy in the free world. The spur of competition, as in other spheres, confers a seldom-realised advantage for the consumer in that he/she is able to purchase and enjoy a very advanced and sophisticated VTR, bristling with features, at an artificially low price. The situation has arisen because manufacturers and retailers have played leapfrog amongst themselves, in their respective spheres of features and profit margins, in an effort to promote themselves in the marketplace.

(a)   (b)

(c)   (d)

Figure 26.1   *The contenders: (a) VHS homedeck, Mitsubishi HS-M60V. (b) S-VHS editing deck, Panasonic NV-HS1000. (c) VHS-C camcorder, JVC GR-AX400. (d) Hi-8 semi-pro camcorder, Sony CCDVX1E*

Manufacturers have the long-term objective of building their reputations, pet formats and market-share; while retailers, locked in the stranglehold of high street competition, have the simple objective of staying in business by maintaining a high volume of low-profit turnover.

Let us now examine the main features of each format, whose tape cassettes are illustrated in Figures 26.2 and 26.3.

VHS

VHS stand for Video Home System: this format was designed by the JVC company in Japan. It was released in Europe in 1978. VHS enjoyed the backing of one of the leading UK rental companies of those days, Thorn–EMI–Ferguson, and this largely accounts for its fast and deep initial penetration into the UK market, aided by the British habit of renting rather than buying television equipment. Simple mechanics characterised the VHS scheme, with an M-wrap threading system using combined threading posts/tape guides as in Figure 17.6. A tilted head drum also simplifies the deck system by permitting a tape path which is parallel to the plane of the deck service, a layout now adopted in all formats.

Figure 26.2 *Cassette packages and playing time compared. The running times given are for SP operation, and are doubled in LP mode*

Figure 26.3 *Cassette sizes compared. Clockwise from bottom right are Video-8, VHS, VHS-C and an audio Compact Cassette, shown here for size comparison*

Standard VHS boasts the widest video track of any current format (49 microns) which offers best S/N ratio conditions for the video signal. The small size and physical simplicity of VHS deck and tape transport arrangements (especially the small-head variations described in Chapter 17) lent themselves well to portable machine applications, and it was in VHS form that the first 'mobile' VTRs appeared.

VHS-C

A clever adaptation of the standard tape and cassette package is VHS-C (Compact) which permits a small and light VHS comcorder. The cassette is a small (92mm × 59 mm × 23 mm) housing containing up to 1 hour's worth of standard 12.7 mm VHS tape. It fits a small camcorder (Figure 26.1c) weighing less than 1 kg, and incorporating a small head drum, thin direct-drive motors and a solid-state image sensor. Back at home the small cassette is loaded, piggy-back style, into a normal-size adaptor shell for replay or editing in the standard VHS machine.

Long-play

Both camcorders and homebase machines, VHS and Video-8, are available with a dual-speed option. On record they are switchable betweeen standard and slow speed, and during replay can automatically recognise which recording speed was used in any cassette offered to them.

In LP mode the capstan speed is halved – to double the playing time of a standard cassette to a maximum (e.g. VHS E240 tape) of eight hours. Early dual-speed machines had separate head pairs for SP and LP modes; later designs incorporate both SP and LP heads in a single pair of ferrite chips; current practice is the use of the *same* pair of heads in both modes. Here the head width is a compromise; in SP, narrower-than-standard tracks are written, leaving between them a form of guard-band. In LP, each recorded track is wider than required, and its excess width is cut down by the erasing action of the next head sweep – see Figure 17.22 and associated text. VHS-LP video tracks are 24.5 microns wide, and the sound and control tracks are recorded in the normal way at the slower speed. The track patterns, then, represent a 'telescoped-up' version of Figure 17.5 with video tracks and control pulses now occupying half the linear tape space of those of a standard VHS recording.

Performance-wise, LP mode is surprisingly good. Video S/N ratio is slightly impaired due to the narrow video tracks; but sound HF response is dramatically cut, where longitudinal tracks are used. Many non-critical viewers cannot tell the difference between SP and LP on replay, so long as it comes from a good tape in good condition. The maximum eight-hour capability of VHS-LP comes into its own when used in conjunction with multi-event timers, and allows, for instance, eight separate one-hour recordings to be made without human intervention.

Even though LP performance is good, its use in camcorders and homebase machines is not recommended unless the programme will be viewed once (non-critically!) before being discarded.

In VTRs and camcorders which use a single pair of compromise-width heads, (most of them, in practice), the still-frame picture performance, especially in SP mode, is very poor due to the impossibility of keeping a narrow head sweep aligned with a narrow tape track when the scanning angle is changed by the cessation of tape movement: this was explained in Chapter 17 and Figures 17.18 and 17.20.

Figure 26.4 Recording signal spectrum for S-VHS format

S-VHS

Super-VHS is an advanced 'high-band' variant of the established format. It uses a high FM carrier, with deviation (Figure 26.4) from 5.4 to 7 MHz and more 'meat' in the video sidebands. To permit this an advanced videotape formula is used in a cassette of conventional size, shape and running time. A very small video head-gap, and 'fine grain' magnetic tape permits a baseband video frequency response approaching 5 MHz, and on-screen resolution better than 400 lines. Noise ratio is also better than that of the low-band formats.

One of the problems of PAL and NTSC colour encoding is the presence of cross-colour effects, discussed in some detail at the end of Chapter 4. Where the programme is originated in these forms little can be done to prevent cross-colour – it stems from the decoding process at the receiving end. For other applications (home videography, MAC and digital TV transmissions, pre-recorded movies on tape etc.) the S-VHS system has provision for complete separation of Y and C (luminance and chrominance) signals throughout the record and playback processes. Where the TV receiver or monitor is equipped with a suitable input socket (S-terminal) and a large screen, S-VHS format is capable of a picture performance superior to that available from a terrestrial broadcast system.

Video-8

The Video-8 format uses 8 mm-wide tape, from which it gets its name. Unlike the others this system was developed by a consortium of companies and accepted for use by over 127 of the world's major audio/video manufacturers. It was the first domestic video system to use digital audio recording, and the first to be designed from the outset for an alternative mode of operation – as a high-quality digital sound-only recording system. Its other advantages are a tiny cassette (Figure 26.3),

Figure 26.5 Video-8 tape threading: (a) cassette in; (b) first stage; (c) wrap completed. Compare with *Figure 17.6*

facilitating miniaturisation of the equipment, be it portable or homebase type; a flying erase head for good edits, also featured in some VHS equipment; and the exploitation of new tape and head materials and techniques for better performance.

Although offered in homebase form for tabletop use, Video-8 is seen mainly as a camcorder format, whose primary advantages are excellent sound, light weight, and high performance – the Video-8 manufacturing companies are particular experts with lenses and TV image sensors.

The Video-8 tape wrap was shown in Figure 20.9. Although the format specifications do not in any way dictate the mechanical layout of the machine, some Video-8 machines use a combination of M-wrap and U-wrap techniques as shown in Figure 26.5. At a is shown the situation where the casssette has just been lowered onto the deck: the pinch-roller and several tape guides have penetrated the cassette behind the front tape loop. In diagram b the first (M-loading) phase is complete: two guides have drawn a loop of tape away from the cassette, VHS-fashion. At this point the loading ring starts to rotate anticlockwise, further wrapping the tape around the head drum, borne on a guide/post ahead of the pinch roller, and prevented from folding back on itself by further ring-mounted guides, diagram c.

Current machines use an M-wrap system, like that of Figure 17.6, the first phase of which involves the entry of the *cassette* into the machine.

Hi-8

As with VHS, development of the basic Video-8 format led to the production of a high-band variant, with picture resolution exceeding 400 lines under favourable circumstances. This, the Hi-8 variant, has higher FM record frequencies and a greater sideband spread than Video-8, as shown in figure 26.6. It's used with ME (metal evaporated) and high-grade MP (metal powder) tapes. In camcorders, which is the form in which nearly all Hi-8 hardware is produced, Figure 26.1*d*, the lens and CCD image sensor are upgraded to accommodate the higher resolution capability, and the bandwidth of the luminance path widened to suit. As with

372

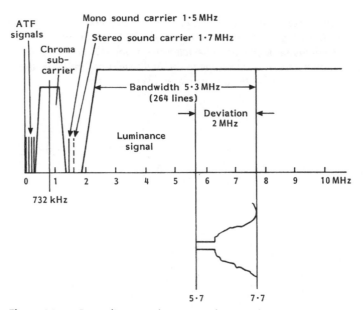

Figure 26.6 *Recording signal spectrum for Hi-8 format*

S-VHS, provision is made for separate handling of Y and C video components throughout the chain, and for their passage through S-links to suitably-equipped TVs, monitors and copy-VTRs.

In other respects, Hi-8 is similar to Video-8 format, with the same cassette dimensions, sound systems, tracking and transport arrangements.

Format comparison

The salient features of the competing formats are shown in Table 26.1 which includes details of some obsolete formats for comparison purposes. The VCR, VCR-LP, Betamax and V2000 formats are really only of academic interest now, but are included to show the progress made over the years.

Regarding performance, too much emphasis should not be put on the details shown in Table 26.1, since it is very difficult, even for those experienced in domestic VCR use, to tell the formats apart by watching replayed pictures from any of them, so long as the comparison is like for like in high-band or low-band. In average domestic conditions, any slight performance differences between the various sytems are sometimes 'swamped' by other shortcomings; a chain is only as good as its weakest link, and many factors beyond the control of the VTR manufacturer can mar performance. A weak aerial signal impairs S/N ratio; worn tape or dirty heads cause excessive drop-out; maladjustment of the tracking control, or a deck in need of servicing leads to mistracking. The TV set in use may

373

Table 26.1 *Formats compared: the main parameters of past and present home formats. All but Video-8 use 12.7 mm ($\frac{1}{2}$ in) tape. The high-band variants S-VHS and Hi-8 have the same physical characteristics as the standard VHS and Video-8 bases shown here. It is interesting to see (right-hand column) how the information density on tape has steadily increased over the years*

	Linear tape speed, cm/sec	Video writing speed, m/sec	Video track width, microns	Azmuth offset of video heads	Head drum diameter, cm	Video track angle to tape	Sound track width, mm (mono or L & R)	Max playing time, hours	Information density on tape (Hrs per sq.metre)
VCR	14.29	8.10	130	±15°	10.5	3.69°	0.7	1	0.16
VCR-LR	6.56	8.10	85	±15°	10.5	3.71°	0.7	2	0.33
VHS	2.34	4.85	49	± 6°	6.2	5.33°	1.0	4‡	0.93
BETA	1.87	5.83	33	± 7°	7.45	5.97°	1.05	3½	1.16
V2000	2.44	5.08	23	±15°	6.5	2.65°	0.65	8*	1.79
Video 8	2.01	3.12	34	±10°	4.0	4.92°	N/A	1½	1.70
VHS-LP	1.17	4.85	24.5	± 6°	6.2	5.35°	1.0	8‡	1.86
V8-LP	1.01	3.12	17	±10°	4.0	4.95°	N/A	3	3.4

*Flip-over cassette: 2 × 4 hours. ‡ 5-hour tapes are available

not be in perfect condition, and maladjustment of the focus control, a worn tube or incorrect tuning can easily halve the definition available from the VTR!

Special features

The provision of extra features has a value which varies greatly from user to user, and in some cases the advantages of these are more apparent than real. All VTRs are capable of recording and replaying TV programmes well, and for the reasons just given, and the fact that signal-processing circuit design does not vary much between machines (leaving out the special case of high-band formats), a £400 model is unlikely to give markedly better performance on normal programme replay than one costing £270. Most domestic VTR use is for record and replay of films, TV serials and sports events, and replay of commercially-recorded cassettes which are purchased or rented, and for these a basic machine is quite adequate.

Most of the extra cost of full-feature VTRs, then, goes into the provision of editing, still-frame, hi-fi stereo sound, Nicam decoders and trick-picture effects using either special head arrangements or field-store memories, with the basic electronics and mechanics largely unchanged. In many households the use made of some of these features and facilities does not justify their cost. Certainly VideoPlus timer programming is generally worthwhile, and good freeze-frame, frame-advance and slow-motion facilities are very useful in educational spheres and for sports enthusiasts. Such things as high-band formats and Nicam hi-fi sound are completely lost if the machine is used with a small-screen mono-sound TV set, however, so careful thought at the time of purchase can save unnecessary expense. The same is true of camcorders. High-band models give a very worth-while increease in definition, again of little use with a small-screen TV, but com-ing into its own with large playback screens and where editing or copying is envisaged. For the latter, too, it's essential to choose a camcorder with an edit-control port if it may ever be used in conjunction with any form of auto-editing system. The value of in-camera 'effects' is somewhat in doubt once the novelty has worn off, but the value of a long zoom range lens and good handling features is never so. Also worth bearing in mind is the audio performance of the two competing camcorder formats, and especially the limited bandwidth and S/N ratio of VHS/VHS-C camcorders which use longitudinal audio tracks, especially in LP mode – this was illustrated in Figure 20.10. VHS hi-fi and all 8 mm systems have better sound, virtually independent of tape speed. Sound apart, there is little *intrinsic* difference in performance between the two formats in their high- and low-band variants, though the 8 mm systems currently offer a wider range of features and types due to the larger number of manufacturers now supporting them.

Index